* 1,000,000 *

Free E-Books

@

www.ForgottenBooks.org

* Alchemy *

The Secret is Revealed

@

www.TheBookofAquarius.com

The Corpuscular Theory of Matter

By

J. J. Thomson

Published by Forgotten Books 2012
Originally Published 1907

PIBN 1000035440

Copyright © 2012 Forgotten Books
www.forgottenbooks.org

BY THE SAME AUTHOR.

THE DISCHARGE OF ELECTRICITY THROUGH GASES.

Crown 8vo. 4/6 net.

ELECTRICITY AND MATTER.

Price 5/- net.

ARCHIBALD CONSTABLE & CO. LTD.

THE CORPUSCULAR
THEORY OF MATTER

BY

J. J. THOMSON, M.A. F.R.S. D.Sc. LL.D. Ph.D.

PROFESSOR OF EXPERIMENTAL PHYSICS, CAMBRIDGE, AND PROFESSOR
OF NATURAL PHILOSOPHY AT THE ROYAL INSTITUTION, LONDON.

NEW YORK
CHARLES SCRIBNER'S SONS
1907

PREFACE

This book is an expansion of a course of lectures given at the Royal Institution in the Spring of 1906. It contains a description of the properties of corpuscles and their application to the explanation of some physical phenomena. In the earlier chapters a considerable amount of attention is devoted to the consideration of the theory that many of the properties of metals are due to the motion of corpuscles diffused throughout the metal. This theory has received strong support from the investigations of Drude and Lorentz; the former has shown that the theory gives an approximately correct value for the ratio of the thermal and electrical conductivities of pure metals and the latter that it accounts for the long-wave radiation from hot bodies. I give reasons for thinking that the theory in its usual form requires the presence of so many corpuscles that their specific heat would exceed the actual specific heat of the metal. I have proposed a modification of the theory which is not open to this objection and which makes the ratio of the conductivities and the long-wave radiation of the right magnitude.

The later chapters contain a discussion of the properties of an atom built up of corpuscles and of positive electricity, the positive electricity being supposed to occupy a much larger volume than the corpuscles. The properties of an atom of this kind are shown to resemble in many respects those of the atoms of the chemical elements. I think that a theory which enables us to picture a kind of model atom and to interpret chemical and physical results in terms of such model may be useful even though the models are crude, for if we picture to ourselves how the model atom must be behaving in some particular physical or chemical process, we not only gain a very vivid conception of the process, but also often suggestions that the process under consideration must be connected with other processes, and thus further investigations are promoted by this method; it also has the advantage of emphasising the unity of chemical and electrical action.

In Chapter VII. I give reasons for thinking that the number of corpuscles in an atom of an element is not greatly in excess of the atomic weight of the element, thus in particular that the number of corpuscles in an atom of hydrogen is not large. Some writers seem to think that this makes the conception of the model atom more difficult. I am unable to follow this view; it seems to me to make the conception easier, since it makes the number of possible atoms much more nearly equal to the number of the chemical elements. It has, however, an important bearing on our conception of the origin of the mass of the atom, as if the number of corpuscles in the atom is of the same order as the atomic weight we cannot regard the mass of an atom as mainly or even appreciably due to the mass of the corpuscles.

I am indebted to Mr. G. W. C. Kaye for assisting in revising the proof sheets.

J. J. THOMSON.

CAMBRIDGE,
July 15, 1907.

CONTENTS

CHAP.		PAGE
I.	Introduction—Corpuscles in Vacuum Tubes	1
II.	The Origin of the Mass of the Corpuscle	28
III.	Properties of a Corpuscle	43
IV.	Corpuscular Theory of Metallic Conduction	49
V.	The Second Theory of Electrical Conduction	86
VI.	The Arrangement of Corpuscles in the Atom	103
VII.	On the Number of Corpuscles in an Atom	142

INDEX 169

THE CORPUSCULAR THEORY OF MATTER

CHAPTER I.

The theory of the constitution of matter which I propose to discuss in these lectures, is one which supposes that the various properties of matter may be regarded as arising from electrical effects. The basis of the theory is electricity, and its object is to construct a model atom, made up of specified arrangements of positive and negative electricity, which shall imitate as far as possible the properties of the real atom. We shall postulate that the attractions and repulsions between the electrical charges in the atom follow the familiar law of the inverse square of the distance, though, of course, we have only direct experimental proof of this law when the magnitude of the charges and the distances between them are enormously greater than those which can occur in the atom. We shall not attempt to go behind these forces and discuss the mechanism by which they might be produced. The theory is not an ultimate one; its object is physical rather than metaphysical. From the point of view of the physicist, a theory of matter is a policy rather than a creed; its object is to connect or co-ordinate apparently diverse phenomena, and above all to suggest, stimulate and direct experiment. It ought to furnish a compass which, if followed, will lead the observer further and further into previously unexplored regions.

Whether these regions will be barren or fertile experience alone will decide; but, at any rate, one who is guided in this way will travel onward in a definite direction, and will not wander aimlessly to and fro.

The corpuscular theory of matter with its assumptions of electrical charges and the forces between them is not nearly so fundamental as the vortex atom theory of matter, in which all that is postulated is an incompressible, frictionless liquid possessing inertia and capable of transmitting pressure. On this theory the difference between matter and non-matter and between one kind of matter and another is a difference between the kinds of motion in the incompressible liquid at various places, matter being those portions of the liquid in which there is vortex motion. The simplicity of the assumptions of the vortex atom theory are, however, somewhat dearly purchased at the cost of the mathematical difficulties which are met with in its development; and for many purposes a theory whose consequences are easily followed is preferable to one which is more fundamental but also more unwieldy. We shall, however, often have occasion to avail ourselves of the analogy which exists between the properties of lines of electric force in the electric field and lines of vortex motion in an incompressible fluid.

To return to the corpuscular theory. This theory, as I have said, supposes that the atom is made up of positive and negative electricity. A distinctive feature of this theory—the one from which it derives its name—is the peculiar way in which the negative electricity occurs both in the atom and when free from matter. We suppose that the negative electricity always occurs as exceedingly fine particles called corpuscles, and that all these corpuscles, whenever they occur, are always of the same size and always carry the same quantity of electricity. Whatever may prove to be the constitution of the atom, we have direct experimental proof of the existence of these corpuscles, and I will begin the discussion of the corpuscular theory with a description of the discovery and properties of corpuscles.

Corpuscles in Vacuum Tubes.

The first place in which corpuscles were detected was a highly exhausted tube through which an electric discharge was passing. When I send an electric discharge through this highly exhausted tube you will notice that the sides of the tube glow with a vivid green phosphorescence. That this is due to something proceeding in straight lines from the cathode—the electrode where the negative electricity enters the tube—can be shown in the following way: the experiment is one made many years ago by Sir William Crookes. A Maltese cross made of thin mica is placed between the cathode and the walls of the tube. You will notice that when I send the discharge through the tube, the green phosphorescence does not now extend all over the end of the tube as it did in the tube without the cross. There is a well-defined cross in which there is no phosphorescence at the end of the tube; the mica cross has thrown a shadow on the tube, and the shape of the shadow proves that the phosphorescence is due to something, travelling from the cathode in straight lines, which is stopped by a thin plate of mica. The green phosphorescence is caused by cathode rays, and at one time there was a keen controversy as to the nature of these rays. Two views were prevalent, one, which was chiefly supported by English physicists, was that the rays are negatively electrified bodies shot off from the cathode with great velocity; the other view, which was held by the great majority of German physicists, was that the rays are some kind of ethereal vibrations or waves.

The arguments in favour of the rays being negatively charged particles are (1) that they are deflected by a magnet in just the same way as moving negatively electrified particles. We know that such particles when a magnet is placed near them are acted upon by a force whose direction is at right angles to the magnetic force, and also at right angles to the direction in which the particles are moving. Thus, if the particles are moving horizontally from east to west, and the magnetic force is horizontal and from north to south, the force acting on the negatively electrified particles will be vertical and downwards.

When the magnet is placed so that the magnetic force is along the direction in which the particle is moving the latter will not be affected by the magnet. By placing the magnet in suitable positions I can show you that the cathode particles move in the way indicated by the theory. The observations that can be made in lecture are necessarily very rough and incomplete; but I may add that elaborate and accurate measurements of the movement of

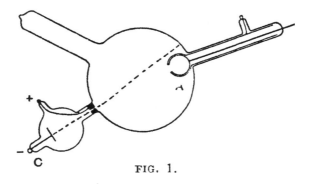

FIG. 1.

cathode rays under magnetic forces have shown that in this respect the rays behave exactly as if they were moving electrified particles.

The next step made in the proof that the rays are negatively charged particles, was to show that when they are caught in a metal vessel they give up to it a charge of negative electricity. This was first done by Perrin. I have here a modification of his experiment. A is a metal cylinder with a hole in it. It is placed so as to be out of the way of the rays coming from C, unless they are deflected by a magnet, and is connected with an electroscope. You see that when the rays do not pass through the hole in the cylinder the electroscope does not receive a charge. I now, by means of a magnet, deflect the rays so that they pass through the hole in the cylinder. You see by the divergence

CORPUSCLES IN VACUUM TUBES.

of the gold-leaves that the electroscope is charged, and on testing the sign of the charge we find that it is negative.

DEFLECTION OF THE RAYS BY A CHARGED BODY.

If the rays are charged with negative electricity they ought to be deflected by an electrified body as well as by a magnet. In the earlier experiments made on this point no such deflection was observed. The reason of this has been shown to be that when the cathode rays pass through a gas they make it a conductor of electricity, so that if there is any appreciable quantity of gas in the vessel through

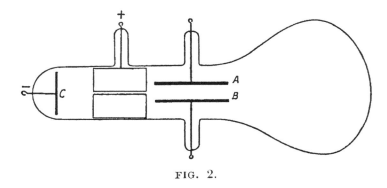

FIG. 2.

which the rays are passing, this gas will become a conductor of electricity, and the rays will be surrounded by a conductor which will screen them from the effects of electric force just as the metal covering of an electroscope screens off all external electric effects. By exhausting the vacuum tube until there was only an exceedingly small quantity of air left in to be made a conductor, I was able to get rid of this effect and to obtain the electric deflection of the cathode rays. The arrangement I used for this purpose is shown in Fig. 2. The rays on their way through the tube pass between two parallel plates, A, B, which can be connected with the poles of a battery of storage cells. The pressure in the tube is very low. You will notice that the rays are very considerably deflected when I connect the plates with the poles of the battery, and that the direction

6 THE CORPUSCULAR THEORY OF MATTER.

of the deflection shows that the rays are negatively charged.

We can also show the effect of magnetic and electric force on these rays if we avail ourselves of the discovery made by Wehnelt, that lime when raised to a red heat emits when negatively charged large quantities of cathode rays. I have here a tube whose cathode is a strip of platinum on which there is a speck of lime. When the piece of platinum is made very hot, a potential difference of 100 volts or so is sufficient to make a stream of cathode rays start from this speck; you will be able to trace the course of the rays by the luminosity they produce as they pass through the gas.

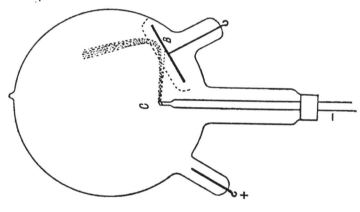

FIG. 3.

You can see the rays as a thin line of bluish light coming from a point on the cathode; on bringing a magnet near it the line becomes curved, and I can bend it into a circle or a spiral, and make it turn round and go right behind the cathode from which it started. This arrangement shows in a very striking way the magnetic deflection of the rays. To show the electrostatic deflection I use the tube shown in Fig. 3. I charge up the plate B negatively so that it repels the pencil of rays which approach it from the spot of lime on the cathode, C. You see that the pencil of rays is deflected from the plate and pursues a curved path whose distance from the plate I can increase or diminish by increasing or diminishing the negative charge on the plate.

We have seen that the cathode rays behave under every test that we have applied as if they are negatively electrified particles; we have seen that they carry a negative charge of electricity and are deflected by electric and magnetic forces just as negatively electrified particles would be.

Hertz showed, however, that the cathode particles possess another property which seemed inconsistent with the idea that they are particles of matter, for he found that they were able to penetrate very thin sheets of metal, for example, pieces of gold-leaf placed between them and the glass, and produce appreciable luminosity on the glass after doing so. The idea of particles as large as the molecules of a gas passing through a solid plate was a somewhat startling

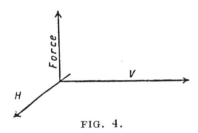

FIG. 4.

one in an age which knew not radium—which does project particles of this size through pieces of metal much thicker than gold-leaf—and this led me to investigate more closely the nature of the particles which form the cathode rays.

The principle of the method used is as follows: When a particle carrying a charge e is moving with the velocity v across the lines of force in a magnetic field, placed so that the lines of magnetic force are at right angles to the motion of the particle, then if H is the magnetic force, the moving particle will be acted on by a force equal to Hev. This force acts in the direction which is at right angles to the magnetic force and to the direction of motion of the particle, so that if the particle is moving horizontally as in the figure and the magnetic force is at right angles to the plane of the paper and towards the reader, then the negatively electrified particle will be acted on by a vertical and upward force. The pencil of rays will therefore be deflected upwards and with it the patch of green phosphorescence where it strikes the walls of the tube. Let now the two parallel plates A and B (Fig. 2) between which the pencil of rays is moving be charged with electricity so that the upper plate is negatively and the lower plate positively electrified, the cathode rays will be repelled from the upper plate with a force Xe where X is the electric force between the plates. Thus, if the plates are charged when the magnetic field is acting on the rays, the magnetic force will tend to send the rays upwards, while the charge on the plates will tend to send them downwards. We can adjust the electric and magnetic forces until they balance and the pencil of rays passes horizontally in a straight line between the plates, the green patch of phosphorescence being undisturbed. When this is the case, the force $He\,v$ due to the magnetic field is equal to Xe—the force due to the electric field—and we have

$$He\,v = Xe$$

$$\text{or} \quad v = \frac{X}{H}$$

Thus, if we measure, as we can without difficulty, the values of X and H when the rays are not deflected, we can determine the value of v, the velocity of the particles. The velocity of the rays found in this way is very great; it varies largely with the pressure of the gas left in the tube. In a very highly exhausted tube it may be 1/3 the velocity of light or about 60,000 miles per second; in tubes not so highly exhausted it may not be more than 5,000 miles per second, but in all cases when the cathode rays are produced in tubes their velocity is much greater than the velocity of any other moving body with which we are acquainted. It is, for example, many thousand times the average velocity with which the molecules of hydrogen are moving at ordinary temperatures, or indeed at any temperature yet realised.

Determination of e/m.

Having found the velocity of the rays, let us in the preceding experiment take away the magnetic force and leave the rays to the action of the electric force alone. Then the particles forming the rays are acted upon by a constant vertical downward force and the problem is practically that of a bullet projected horizontally with a velocity v and falling under gravity. We know that in time t the body will fall a depth equal to $\frac{1}{2} g t^2$ where g is the vertical acceleration; in our case the vertical acceleration is equal to $X e/m$ where m is the mass of the particle, the time it is falling is l/v where l is the length of path measured horizontally, and v the velocity of projection. Thus, the depth the particle has fallen when it reaches the glass, i.e., the downward displacement of the patch of phosphorescence where the rays strike the glass, is equal to

$$\frac{1}{2} \frac{X e\, l^2}{m\, v^2}$$

We can easily measure d the distance the phosphorescent patch is lowered, and as we have found v and X and l are easily measured, we can find e/m from the equation:

$$\frac{e}{m} = \frac{2 d\, v^2}{X\, l^2}$$

The results of the determinations of the values of e/m made by this method are very interesting, for it is found that however the cathode rays are produced we always get the same value of e/m for all the particles in the rays. We may, for example, by altering the shape of the discharge tube and the pressure of the gas in the tube, produce great changes in the velocity of the particles, but unless the velocity of the particles becomes so great that they are moving nearly as fast as light, when, as we shall see, other considerations have to be taken into account, the value of e/m is constant. The value of e/m is not merely independent of the velocity. What is even more remarkable is that it is independent of the kind of electrodes we use and also of the kind of gas in the tube. The particles which form the cathode rays must come either from the gas in the tube or from the electrodes; we may, however, use any kind of substance we please for the electrodes and fill the tube with gas of any kind, and yet the value of e/m will remain unaltered.

This constant value is, when we measure e/m in the C. G. S. system of magnetic units, equal to about 1.7×10^7. If we compare this with the value of the ratio of the mass to the charge of electricity carried by any system previously known, we find that it is of quite a different order of magnitude. Before the cathode rays were investigated the charged atom of hydrogen met with in the electrolysis of liquids was the system which had the greatest known value for e/m, and in this case the value is only 10^4; hence for the corpuscle in the cathode rays the value of e/m is 1,700 times the value of the corresponding quantity for the charged hydrogen atom. This discrepancy must arise in one or other of two ways, either the mass of the corpuscle must be very small compared with that of the atom of hydrogen, which until quite recently was the smallest mass recognised in physics, or else the charge on the corpuscle must be very much greater than that on the hydrogen atom. Now it has been shown by a method which I shall shortly describe that the electric charge is practically the same in the two cases; hence we are driven to the conclusion that the mass of the corpuscle is only about 1/1700 of that of the hydrogen atom. Thus the atom is not the ultimate limit to the subdivision of matter; we may go further and get to the corpuscle, and at this stage the corpuscle is the same from whatever source it may be derived.

Corpuscles very widely distributed.

It is not only from what may be regarded as a somewhat artificial and sophisticated source, viz., cathode rays, that we can obtain corpuscles. When once they had been discovered it was found that they were of very general occurrence. They are given out by metals when raised to

CORPUSCLES IN VACUUM TUBES.

a red heat: you have already seen what a copious supply is given out by hot lime. Any substance when heated gives out corpuscles to some extent; indeed, we can detect the emission of them from some substances, such as rubidium and the alloy of sodium and potassium, even when they are cold; and it is perhaps allowable to suppose that there is some emission by all substances, though our instruments are not at present sufficiently delicate to detect it unless it is unusually large.

Corpuscles are also given out by metals and other bodies, but especially by the alkali metals, when these are exposed to light. They are being continually given out in large quantities, and with very great velocities by radio-active substances such as uranium and radium; they are produced in large quantities when salts are put into flames, and there is good reason to suppose that corpuscles reach us from the sun.

The corpuscle is thus very widely distributed, but wherever it is found it preserves its individuality, e/m being always equal to a certain constant value.

The corpuscle appears to form a part of all kinds of matter under the most diverse conditions; it seems natural, therefore, to regard it as one of the bricks of which atoms are built up.

MAGNITUDE OF THE ELECTRIC CHARGE CARRIED BY THE CORPUSCLE.

I shall now return to the proof that the very large value of e/m for the corpuscle as compared with that for the atom of hydrogen is due to the smallness of m the mass, and not to the greatness of e the charge. We can do this by actually measuring the value of e, availing ourselves for this purpose of a discovery by C. T. R. Wilson, that a charged particle acts as a nucleus round which water vapour condenses, and forms drops of water. If we have air saturated with water vapour and cool it so that it would be supersaturated if there were no deposition of moisture, we know that if any dust is present, the particles of dust act

12 THE CORPUSCULAR THEORY OF MATTER.

as nuclei round which the water condenses and we get the too familiar phenomena of fog and rain. If the air is quite dust-free we can, however, cool it very considerably without any deposition of moisture taking place. If there is no dust, C. T. R. Wilson has shown that the cloud does not form until the temperature has been lowered to such a point that the supersaturation is about eightfold. When, however, this temperature is reached, a thick fog forms, even in dust-free air. When charged particles are present

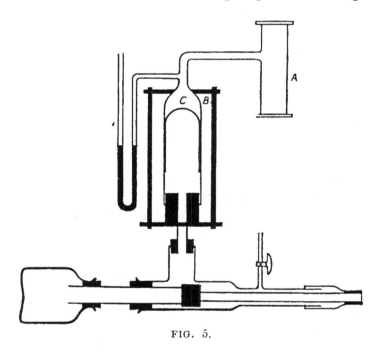

FIG. 5.

in the gas, Wilson showed that a much smaller amount of cooling is sufficient to produce the fog, a fourfold supersaturation being all that is required when the charged particles are those which occur in a gas when it is in the state in which it conducts electricity. Each of the charged particles becomes the centre round which a drop of water forms; the drops form a cloud, and thus the charged particles, however small to begin with, now become visible and can be observed. The effect of the charged particles on the formation of a cloud can be shown very distinctly by the following experiment. The vessel A, which is in contact with water, is saturated with moisture at the temperature of the room. This vessel is in communication with B, a cylinder in which a large piston, C, slides up and down; the piston, to begin with, is at the top of its travel; then by suddenly exhausting the air from below the piston, the pressure of the air above it will force it down with great rapidity, and the air in the vessel A will expand very quickly. When, however, air expands it gets cool; thus the air in A gets colder, and as it was saturated with moisture before cooling, it is now supersaturated. If there is no dust present, no deposition of moisture will take place unless the air in A is cooled to such a low temperature that the amount of moisture required to saturate it is only about 1/8 of that actually present. Now the amount of cooling, and therefore of supersaturation, depends upon the travel of the piston; the greater the travel the greater the cooling. I can regulate this travel so that the supersaturation is less than eightfold, and greater than fourfold. We now free the air from dust by forming cloud after cloud in the dusty air, as the clouds fall they carry the dust down with them, just as in nature the air is cleared by showers. We find at last that when we make the expansion no cloud is visible. We now put the gas in a conducting state by bringing a little radium near the vessel A; this fills the gas with large quantities of both positively and negatively electrified particles. On making the expansion now, an exceedingly dense cloud is formed. That this is due to the electrification in the gas can be shown by the following experiment: Along the inside walls of the vessel A we have two vertical insulated plates which can be electrified; if these plates are electrified they will drag the charged particles out of the gas as fast as they are formed, so that by electrifying the plates we can get rid of, or at any rate largely reduce, the number of electrified particles in the gas. I now repeat the experiment, electrifying the plates before bringing up the radium. You see that the presence of the radium hardly increases the small amount of cloud. I now discharge the plates, and on making the expansion the cloud is so dense as to be quite opaque.

We can use the drops to find the charge on the particles, for when we know the travel of the piston we can deduce the amount of supersaturation, and hence the amount of water deposited when the cloud forms. The water is deposited in the form of a number of small drops all of the same size; thus the number of drops will be the volume of the water deposited divided by the volume of one of the drops. Hence, if we find the volume of one of the drops we can find the number of drops which are formed round the charged particles. If the particles are not too numerous, each will have a drop round it, and we can thus find the number of electrified particles.

If we observe the rate at which the drops slowly fall down we can determine the size of the drops. In consequence of the viscosity or friction of the air small bodies do not fall with a constantly accelerated velocity, but soon reach a speed which remains uniform for the rest of the fall; the smaller the body the slower this speed, and Sir George Stokes has shown that v, the speed at which a drop of rain falls, is given by the formula—

$$v = \frac{2}{9} \frac{g a^2}{\mu}$$

where a is the radius of the drop, g the acceleration due to gravity, and μ the co-efficient of viscosity of the air. If we substitute the values of g and μ, we get

$$v = 1\cdot 28 \times 10^6 \, a^2$$

Hence, if we measure v we can determine a, the radius of the drop. We can, in this way, find the volume of a drop, and may therefore, as explained above, calculate the number of drops, and therefore the number of electrified particles. It is a simple matter to find, by electrical methods, the total quantity of electricity on these particles; and hence, as we know the number of particles, we can deduce at once the charge on each particle.

This was the method by which I first determined the charge on the particle. H. A. Wilson has since used a simpler method founded on the following principles. C. T. R. Wilson has shown that the drops of water condense more easily on negatively electrified particles than on positively electrified ones. Thus, by adjusting the expansion, it is possible to get drops of water round the negative particles and not round the positive; with this expansion, therefore, all the drops are negatively electrified. The size of these drops, and therefore their weight, can, as before, be determined by measuring the speed at which they fall under gravity. Suppose now, that we hold above the drops a positively electrified body, then since the drops are negatively electrified they will be attracted towards the positive electricity and thus the downward force on the drops will be diminished, and they will not fall so rapidly as they did when free from electrical attraction. If we adjust the electrical attraction so that the upward force on each drop is equal to the weight of the drop, the drops will not fall at all, but will, like Mahomet's coffin, remain suspended between heaven and earth. If, then, we adjust the electrical force until the drops are in equilibrium and neither fall nor rise, we know that the upward force on the drop is equal to the weight of the drop, which we have already determined by measuring the rate of fall when the drop was not exposed to any electrical force. If X is the electrical force, e the charge on the drop, and w its weight, we have, when there is equilibrium—

$$X e = w.$$

Since X can easily be measured, and w is known, we can use this relation to determine e, the charge on the drop. The value of e found by these methods is $3{\cdot}1 \times 10^{-10}$ electrostatic units, or 10^{-20} electromagnetic units. This value is the same as that of the charge carried by a hydrogen atom in the electrolysis of dilute solutions, an approximate value of which has long been known.

It might be objected that the charge measured in the

preceding experiments is the charge on a molecule or collection of molecules of the gas, and not the charge on a corpuscle. This objection does not, however, apply to another form in which I tried the experiment, where the charges on the particles were got, not by exposing the gas to the effects of radium, but by allowing ultra-violet light to fall on a metal plate in contact with the gas. In this case, as experiments made in a very high vacuum show, the electrification which is entirely negative escapes from the metal in the form of corpuscles. When a gas is present, the corpuscles strike against the molecules of the gas and stick to them. Thus, though it is the molecules which are charged, the charge on a molecule is equal to the charge on a corpuscle, and when we determine the charge on the molecules by the methods I have just described, we determine the charge carried by the corpuscle. The value of the charge when the electrification is produced by ultra-violet light is the same as when the electrification is produced by radium.

We have just seen that e, the charge on the corpuscle, is in electromagnetic units, equal to 10^{-20}, and we have previously found that e/m, m being the mass of a corpuscle, is equal to 1.7×10^7, hence $m = 6 \times 10^{-28}$ grammes.

We can realise more easily what this means if we express the mass of the corpuscle in terms of the mass of the atom of hydrogen. We have seen that for the corpuscle $e/m = 1.7 \times 10^7$; while if E is the charge carried by an atom of hydrogen in the electrolysis of dilute solutions, and M the mass of the hydrogen atom, $E/M = 10^4$; hence $e/m = 1700\ E/M$. We have already stated that the value of e found by the preceding methods agrees well with the value of E, which has long been approximately known. Townsend has used a method in which the value of e/E is directly measured and has showed in this way also that e is equal to E; hence, since $e/m = 1700\ E/M$, we have $M = 1700\ m$, i.e., the mass of a corpuscle is only about 1/1700 part of the mass of the hydrogen atom.

In all known cases in which negative electricity occurs in

gases at very low pressures it occurs in the form of corpuscles, small bodies with an invariable charge and mass. The case is entirely different with positive electricity.

The Carriers of Positive Electricity.

We get examples of positively charged particles in various phenomena. One of the first cases to be investigated was that of the "Canalstrahlen" discovered by Goldstein. I have here a highly exhausted tube with a cathode, through which a large number of holes has been bored. When I send a discharge through this tube you will see the cathode rays shooting out in front of the cathode. In addition to these, you see other rays streaming through the holes in the cathode, and travelling through the gas at the back of

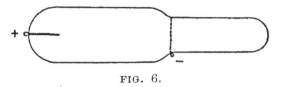

FIG. 6.

the cathode. These are called "Canalstrahlen." You notice that, like the cathode rays, they make the gas luminous as they pass through it, but the colour of the luminosity due to the canalstrahlen is not the same as that due to the cathode rays. The distinction is exceptionally well marked in helium, where the luminosity due to the canalstrahlen is tawny, and that due to the cathode rays bluish. The luminosity, too, produced when the rays strike against a solid is also of quite a different character. This is well shown by allowing both cathode rays and canalstrahlen to strike against lithium chloride. Under the cathode rays the salt gives out a steely blue light, and the spectrum is a continuous one; under the canalstrahlen the salt gives out a brilliant red light, and the spectrum shows the lithium line. It is a very interesting fact that the lines in the spectra of the alkali metals are very much more easily

18 THE CORPUSCULAR THEORY OF MATTER.

obtained when the canalstrahlen fall on salts of the metal than when they fall on the metal itself. Thus when a pool of the liquid alloy of sodium and potassium is bombarded by canalstrahlen the specks of oxide on the surface shine with a bright yellow light, while the untarnished part of the surface is quite dark.

The canalstrahlen are deflected by a magnet, though not to anything like the same extent as the cathode rays. Their deflection, too, is in the opposite direction, showing that they are positively charged.

Value of e/m for the Particles in the Canalstrahlen.

W. Wien has applied the methods described in connection with the cathode rays to determine the value of e/m for the particles in the canalstrahlen. The contrast between the results obtained for the two rays is very interesting. In the case of the cathode rays the velocity of different rays in the same tube may be different, but the value of e/m for these rays is independent of the velocity as well as of the nature of the gas and the electrodes. In the case of the canalstrahlen we get in the same pencil of rays not merely variations in the velocity, but also variations in the value of e/m. The difference between the values of e/m for the cathode rays and the canalstrahlen is also very remarkable. For the cathode rays e/m always equal to 1.7×10^7; while for canalstrahlen the greatest value ever observed is 10^4, which is also the value of e/m for the hydrogen ions in the electrolysis of dilute solutions. When the canalstrahlen pass through hydrogen the value of e/m for a large portion of the rays is 10^4. There are, however, some rays present even in hydrogen, for which e/m is much less than 10^4, and which are but slightly deflected even by very intense magnetic fields. When the canalstrahlen pass through very pure oxygen, Wien found that the value of e/m for the most conspicuous rays was about 750, which is not far from what it would be if the charge were the same as for the canalstrahlen in hydrogen, while the mass was greater in the proportion of the mass of an atom of oxygen to that of an atom of hydrogen. Along with these rays in oxygen there were others having still smaller values of e/m, and some having e/m equal to 10^4.

As the canalstrahlen or rays of positive electricity are a very promising field for investigations on the nature of positive electricity, I have recently made a series of experiments on these rays in different gases, measuring the deflections they experience when exposed to electric and magnetic forces and thus deducing the values of e/m and v. I find, when the pressure of the gas is not too low,

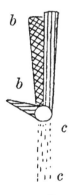

FIG. 7.

The portions with the cross shading is the deflection under both electric and magnetic force; the portion with vertical shading the deflection under magnetic force; that with the horizontal shading under electric force alone.

that the bright spot produced by the impact of these rays on the phosphorescent screen is deflected by electric and magnetic forces into a continuous band extending, as shown in Fig. 7, on both sides of the undeflected portion, the portion on one side (cc) is very much fainter than that on the other, and also somewhat shorter. The direction of the deflection of the band cc shows that it is produced by particles charged with *negative* electricity, while the brighter band bb is due to particles charged with positive electricity. The negatively charged particles which produce the band cc are not corpuscles, for from the deflections in this band we can find the value of e/m; as this value comes out of the order 10^4, we see that the mass of the carrier is comparable with that of an atom, and therefore immensely greater than that of a corpuscle. When the pressure is very low the portion of the phosphorescence deflected in the negative direction disappears and the phosphorescent spot, instead of being stretched by the electric and magnetic forces into a continuous band, is broken up into two patches, as in the curved parts of Figs. 8 and 9. Fig. 8 is the appearance at exceedingly low pressures, Fig. 9 that at a somewhat higher pressure. For one of these patches the maximum value of e/m is about 10^4, and for the other about 5×10^3. The appearance of the patches and the values of e/m at these very low pressures are the same whether

FIG. 8.
The curved patches represent the deflection under both electric and magnetic force.

FIG. 9.

the tube is filled originally with air, hydrogen, or helium. Another experiment I tried was to exhaust the tube until the pressure was too low for the discharge to pass, and then to introduce into the tube a very small quantity of gas, this increases the pressure and the discharge is able to pass through the tube. The following gases were admitted into the tube: air, carbonic oxide, oxygen, hydrogen, helium, argon and neon, but whatever the gas the appearance of the phosphorescence was the same. In every case there were two patches, one having $e/m = 10^4$, the other $e/m = 5 \times 10^3$. At these very low pressures the intensity of the electric field in the discharge tube is very great.

When the pressure in the tube is not very low the nature of the positive rays depends to a very considerable extent upon the kind of gas with which the tube is filled. Thus, for example, in air at these pressures the phosphorescent spot is stretched out into a straight band as in Fig. 7; the maximum value of e/m for this band is 10^4. In hydrogen at suitable pressures we get the spot stretched out into two bands as in Fig. 10; for one of these bands the maximum

FIG. 10.

value of e/m is 10^4, while for the other it is 5×10^3. In helium we also get two bands as in Fig. 11, but while the maximum value of e/m in one of these bands is 10^4, the same as for the corresponding band in hydrogen, the maximum value of e/m in the other band is only $2·5 \times 10^3$. We see from this that the ratio of the masses of the carriers

FIG. 11.

in the two bands is equal to the ratio of the masses of the atoms of hydrogen and helium.

At some pressures we get three bands in helium, the value of e/m being respectively 10^4, 5×10^3, and $2·5 \times 10^3$.

The continuous band into which the bright phosphorescent spot is stretched out when the pressure is not exceedingly low can be explained as follows:—

The rays on their way to the screen have to pass through gas which is ionised by the passage through it of the rays; this gas will therefore contain free corpuscles. The particles which constitute the rays start with a charge of positive electricity; some of these in their journey through the gas may attract a corpuscle, the negative charge on which will neutralise the positive charge on the particle. The particles when in this neutral state may be ionised by collision and reacquire a positive charge, or by attracting another corpuscle they may become negatively charged, and this process may be repeated several times in their journey to the screen. Thus, some of the particles, instead of being positively charged for the whole of the time they are exposed to electric and magnetic forces, may be for a part of that time without a charge or even have a negative charge. Now the deflection of a particle will be proportional to the average value of its charge while under the action of electric and magnetic forces; if the particle is without charge for a part of the time, its deflection will be less than that of a particle which has retained its positive charge for the whole of the journey, while the small number of particles, which have a negative charge for a longer time than they have a positive, will be deflected in the opposite direction and produce the faint tail of phosphorescence which is deflected in the opposite direction to the main portion.

It is remarkable and suggestive that even when great care is taken to eliminate hydrogen from the tube, we get at all pressures a large quantity of rays for which e/m is equal to 10^4, the value for the hydrogen atom; and in many cases this is the only definite value of e/m to be observed, for the continuous band in which we have all values of e/m is due, as we have seen, not to changes in m, but to changes in the average value of e.

'If the presence of rays for which $e/m = 10^4$ was entirely due to hydrogen present as an impurity in the gas with which the tube is filled, the positive particles being hydrogen ionised by the corpuscles projected from the cathode, we should have expected, since the ionisation consists in the detachment of a corpuscle from the molecule, that the positively charged particles would be molecules and not atoms of hydrogen.

Again, at very low pressures, when the electric field is very intense, we get the same two types of carriers whatever kind of gas is in the tube. For one of these types $e/m = 10^4$ and for the other $e/m = 5 \times 10^3$; the second value corresponds to the positive particles which are given out by radio-active substances. The most obvious interpretation of this result is that under the conditions existing in the discharge tube at these very low pressures all gases give off positive particles which resemble corpuscles, in so far as they are independent of the nature of the gas from which they are derived, but which differ from the corpuscles in having masses comparable with the mass of an atom of hydrogen, while the mass of a corpuscle is only 1/1700 of this mass. One type of positive particle has a mass equal to that of an atom of hydrogen, the other type has a mass double this; and the experiments I have just described indicate that when the pressure is very low and the electric field very intense, all the positively electrified particles are of one or other of these types.

We have seen that for the positively charged particles in the canalstrahlen the value of e/m depends, when the pressure is not too low, on the kind of gas in the tube, and is such that the least value of m is comparable with the mass of an atom of hydrogen, and is thus always immensely greater than the carriers of the negative charge in the cathode rays. We know of no case where the mass of the positively charged particle is less than that of an atom of hydrogen.

Positive Ions from Hot Wires.

When a metallic wire is raised to a red heat it gives out positively electrified particles. I have investigated the values of e/m for these particles, and find that they show the same peculiarities as the positively charged particles in the canalstrahlen. The particles given off by the wire are not all alike. Some have one value of e/m, others another,

24 THE CORPUSCULAR THEORY OF MATTER.

but the greatest value I found in my experiments where the wire was surrounded by air at a low pressure was 720, and there were many particles for which e/m was very much smaller, and which were hardly affected even by very strong magnetic fields.

Positive Ions from Radio-active Substances.

The various radio-active substances, such as radium, polonium, uranium, and actinium, shoot out with great velocity positively electrified particles which are called α rays. The values of e/m for these particles have been measured by Rutherford, Des Coudres, Mackenzie, and Huff, and for all the substances hitherto examined—radium and its transformation products, polonium, and actinium—the value of e/m is the same and equal to 5×10^3, the same as for one type of ray in the vacuum tube. The velocity with which the particles move varies considerably from one substance to another. As these substances all give off helium, there is *primâ facie* evidence that the α particles are helium. For a helium atom with a single charge, e/m is $2\cdot5 \times 10^3$, hence if the α particles are helium atoms they must carry a double charge; the large value of e/m shows that the carriers of the positive charge must be atoms, or molecules of some substance with a small atomic weight. Hydrogen and helium are the only substances with an atomic weight small enough to be compatible with so large a value of e/m as 5,000, and of these, helium is known to be given off by radio-active substances, whereas we have as yet no evidence that there is any evolution of hydrogen.

Positive particles having $e/m = 5 \times 10^3$ are found, as we have seen, in all vacuum tubes carrying an electric discharge when the pressure in the tube is very low; the velocity of these particles is very much less than that of the α particles. From the researches of Bragg, Kleeman, and Rutherford, it appears that the α particles lose their power of ionisation and of producing phosphorescence when their velocity is reduced by passing through absorbing substances to about 10^9 cm/sec. The

CORPUSCLES IN VACUUM TUBES. 25

interesting point about this result is that the positively electrified particles in a discharge tube can produce ionisation and phosphorescence when their velocity is very much smaller than this.

This may possibly be due to the α particles being much fewer in number than the positively charged particles in a discharge tube ; and that as the α particles are so few and far between, a particle in its attempts at ionisation or at producing phosphorescence receives no assistance from its companions. Thus, if ionisation or phosphorescence requires a certain amount of energy to be communicated to a system, all that energy has to come from one particle. When, however, as in a discharge tube, the stream of particles is much more concentrated, the energy required by the system may be derived from more than one particle, the energy given to the system by one particle not having been entirely lost before additional energy is supplied by another particle. Thus the effects produced by the particles might be cumulative and the system might ultimately receive the required amount of energy by contributions from several particles. Thus, although the contribution from any one particle might be insufficient to produce ionisation or phosphorescence, the cumulative effects of several might be able to do so.

Another way in which the sudden loss of ionising power might occur is that the power of producing ionisation may be dependent on the possession of an electric charge by the particle, and that when the velocity of the particle falls below a certain value, the particle is no longer able to escape from a negatively charged corpuscle when it passes close to it, but retains the corpuscle as a kind of satellite, the two forming an electrically neutral system, and that inasmuch as the chance of ionisation by collision diminishes as the velocity increases, when the velocity exceeds a certain value, such a neutral system is not so likely to be ionised and again acquire a charge of electricity as the more slowly moving particles in a discharge tube.

These investigations on the properties of the carriers of

positive electricity prove: (1) that whereas in gases at very low pressures the carriers of negative electricity have an exceedingly small mass, only about 1/1700 of that of the hydrogen atom, the mass of the carriers of positive electricity is never less than that of the hydrogen atom; (2) that while the carrier of negative electricity, the corpuscle, has the same mass from whatever source it may be derived, the mass of the carrier of the positive charge may be variable: thus in hydrogen the smallest of the positive particles seems to be the hydrogen atom, while in helium, at not too low a pressure, the carrier of the positive electricity is partly, at any rate, the helium atom. All the evidence at our disposal shows that even in gases at the lowest pressures the positive electricity is always carried by bodies at least as large as atoms; the negative electricity, on the other hand, is under the same circumstances carried by corpuscles, bodies with a constant and exceedingly small mass.

The simplest interpretation of these results is that the positive ions are the atoms or groups of atoms of various elements from which one or more corpuscles have been removed. That, in fact, the corpuscles are the vehicles by which electricity is carried from one body to another, a positively electrified body differing from the same body when unelectrified in having lost some of its corpuscles while the negative electrified body is one with more corpuscles than the unelectrified one.

In the old one-fluid theory of electricity, positive or negative electrification was due to an excess or deficiency of an "electric fluid." On the view we are considering positive or negative electrification is due to a defect or excess in the number of corpuscles. The two views have much in common if we suppose that the "electric fluid" is built up of corpuscles.

In the corpuscular theory of matter we suppose that the atoms of the elements are made up of positive and negative electricity, the negative electricity occurring in the form of corpuscles. In an unelectrified atom there are as many units of positive electricity as there are of negative; an atom with a unit positive charge is a neutral atom which has lost one corpuscle, while an atom with a unit negative charge is a neutral atom to which an additional corpuscle has been attached. No positively electrified body has yet been found with a mass less than that of a hydrogen atom. We cannot, however, without further investigation infer from this that the mass of the unit charge of positive electricity is equal to the mass of the hydrogen atom, for all we know about the electrified system is, that the positive electricity is in excess by one unit over the negative electricity; any system containing n units of positive electricity and $(n-1)$ corpuscles would satisfy this condition whatever might be the value of n. Before we can deduce any conclusions as to the mass of the unit of positive electricity we must know something about the number of corpuscles in the system. We shall give, later on, methods by which we can obtain this information; we may, however, state here that these methods indicate that the number of corpuscles in an atom of any element is proportional to the atomic weight of the element —it is a multiple, and not a large one, of the atomic weight of the element. If this result is right, there cannot be a large number of corpuscles and therefore of units of positive electricity in an atom of hydrogen, and as the mass of a corpuscle is very small compared with that of an atom of hydrogen, it follows that only a small fraction of the mass of the atom can be due to the corpuscle. The bulk of the mass must be due to the positive electricity, and therefore the mass of unit positive charge must be large compared with that of the corpuscle—the unit negative charge.

From the experiments described on p. 19 we conclude that positive electricity is made up of units, which are independent of the nature of the substance which is the seat of the electrification.

CHAPTER II.

THE ORIGIN OF THE MASS OF THE CORPUSCLE.

THE origin of the mass of the corpuscle is very interesting, for it has been shown that this mass arises entirely from the charge of electricity on the corpuscle. We can see how this comes about in the following way. If I take an uncharged body of mass M at rest and set it moving with the velocity V, the work I shall have to do on the body is equal to the kinetic energy it has acquired, i.e., to $\frac{1}{2} MV^2$. If, however, the body is charged with electricity I shall have to do more work to set it moving with the same velocity, for a moving charged body produces magnetic force, it is surrounded by a magnetic field

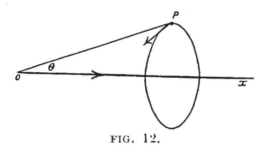

FIG. 12.

and this field contains energy; thus when I set the body in motion I have to supply the energy for this magnetic as well as for the kinetic energy of the body. If the charged body is moving along the line OX, the magnetic force at a point P is at right angles to the plane POX; thus the lines of magnetic forces are circles having OX for their axis. The magnitude of the force at P is equal to $\frac{e V \sin \theta}{r^2}$ where θ denotes the angle POX. Now in a magnetic field the energy per unit volume at any place where the magnetic force is

ORIGIN OF THE MASS OF THE CORPUSCLE.

equal to H is $H^2/8\pi$. Thus the energy per unit volume at P arising from the magnetic force produced by the moving charge is $\dfrac{1}{8\pi}\dfrac{e^2 V^2 \sin^2\theta}{OP^4}$, and by taking the sum of the energy throughout the volume surrounding the charge, we find the amount of energy in the magnetic field. If the moving body is a conducting sphere of radius a, a simple calculation shows that the energy in the magnetic field is equal to $\dfrac{1}{3}\dfrac{e^2 V^2}{a}$. The energy which has to be supplied to set the sphere in motion is this energy *plus* the kinetic energy of the sphere, *i.e.*, it is equal to

$$\tfrac{1}{2} m V^2 + \tfrac{1}{3}\dfrac{e^2}{a} V^2$$

or $$\tfrac{1}{2}\left(m + \tfrac{2}{3}\dfrac{e^2}{a}\right) V^2.$$

Thus the energy is the same as if it were the kinetic energy of a sphere with a mass $m + \dfrac{2}{3}\dfrac{e^2}{a}$ instead of m. Thus the apparent mass of the electrified body is not m but $m + \dfrac{2}{3}\dfrac{e^2}{a}$. The seat of this increase in mass is not in the electrified body itself but in the space around it, just as if the ether in that space were set in motion by the passage through it of the lines of force proceeding from the charged body, and that the increase in the mass of the charged body arose from the mass of the ether set in motion by the lines of electric force. It may make the consideration of this increase in mass clearer if we take a case which is not electrical but in which an increase in the apparent mass occurs from causes which are easily understood. Suppose that we start a sphere of mass M with a velocity V in a vacuum, the work which has to be done on the sphere is $\tfrac{1}{2} M V^2$. Let us now immerse the sphere in water: the work required to start the sphere with the same velocity will evidently be greater than when it was in the vacuum, for the motion of the sphere will set the water around it in motion. The water will have kinetic energy, and this, as well as the kinetic energy of the sphere, has to be supplied when the sphere is moved. It has been shown by Sir George Stokes that the energy in the water is equal to $\frac{1}{2} M_1 V^2$ where M_1 is the mass of half the volume of the water displaced by the sphere. Thus the energy required to start the sphere is $\frac{1}{2} (M + M_1) V^2$, and the sphere behaves as if its mass were $M + M_1$ and not M, and for many purposes we could neglect the effect of the water if we supposed the mass of the sphere to be increased in the way indicated. If we suppose the lines of electric force proceeding from the charged body to set the ether in motion and assume the ether has mass, then the origin of the increase of mass arising from electrification would be very analogous to the case just considered. The increase in mass due to the charge is $\frac{2}{3} \frac{e^2}{a}$; thus for a given charge the increase in the mass is greater for a small body than for a large one. Now for bodies of ordinary size this increase of mass due to electrification is for any realisable charges quite insignificant in comparison with the ordinary mass. But since this addition to the mass increases rapidly as the body gets smaller, the question arises, whether in the case of these charged and exceedingly small corpuscles the electrical mass, as we may call it, may not be quite appreciable in comparison with the other (mechanical) mass. We shall now show that this is the case; indeed for corpuscles there is no other mass: all the mass is electrical.

The method by which this result has been arrived at is as follows: The distribution of magnetic force near a moving electrified particle depends upon the velocity of the particle, and when the velocity approaches that of light, is of quite a different character from that near a slowly moving particle. Perhaps the clearest way of seeing this is to follow the changes which occur in the distribution of the electric force round a charged body as its velocity is gradually increased. When the body is at rest the electric

ORIGIN OF THE MASS OF THE CORPUSCLE. 31

force is uniformly distributed round the body, *i.e.*, as long as we keep at the same distance from the charged body the electric force remains the same whether we are to the east, west, north or south of the particle; the lines of force which come from the body spread out uniformly in all directions. When the body is moving this is no longer the case, for if the body is moving along the line OA (Fig. 13), the lines of electric force tend to leave the regions in the neighbourhood of OA and OB, which we shall call the polar regions, and crowd towards a plane drawn through O at right angles to OA; the regions in the neighbourhood

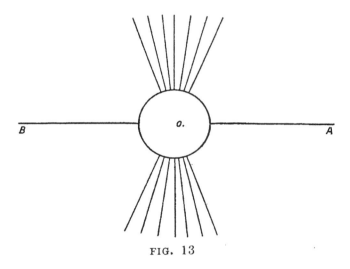

FIG. 13

of this plane we shall call the equatorial regions. This crowding of the lines of force is exceedingly slight when the velocity of the body is only a small fraction of that of light, but it becomes very marked when the velocity of the body is nearly equal to that velocity; and when the body moves at the same speed as light all the lines of force leave the region round OA and crowd into the plane through O at right angles to OA, *i.e.*, the lines of force have swung round until they are all at right angles to the direction in which the particle is moving. The effect of this crowding of the lines of force towards the equatorial plane is to weaken the magnetic force in the polar and increase it in the equatorial regions. The polar regions are those where the magnetic force was originally weak, the equatorial regions those where it was strong. Thus the effect of the crowding is to increase relatively the strength of the field in the strong parts of the field and to weaken it in the weak parts. This makes the energy in the field greater than if there were no crowding, in which case the energy is $\dfrac{1}{3}\dfrac{e^2v^2}{a}$ where e is the charge, v the velocity and a the radius of the sphere. When we allow for the crowding, the energy will be $\dfrac{1}{3}\alpha\dfrac{e^2v^2}{a}$, where α is a quantity which will be equal to unity when v is small compared with c the velocity of light, but becomes very large when v approaches c. The part of the mass arising from the charge is $\dfrac{2}{3}\alpha\dfrac{e^2}{a}$, thus since α depends upon v—the velocity of the particle—the electrical mass will depend upon v, and thus this part of the mass has the peculiarity that it is not constant but depends upon the velocity of the particle. Thus if an appreciable part of the mass of the corpuscle is electrical in origin, the mass of rapidly moving corpuscles will be greater than that of slow ones, while if the mass were in the main mechanical, it would be independent of the velocity. Radium gives out corpuscles which move with velocities comparable with that of light and which are therefore very suitable for testing whether or not this increase in the mass of a corpuscle with its velocity takes place. This test has been applied by Kaufmann, who has measured the value of m/e for the various corpuscles moving with different velocities given out by radium. We can calculate the value of the coefficient α—the quantity which expresses the effect of the velocity on the mass. The value of this quantity depends to some extent on the view we take as to the distribution of electricity on the corpuscle; we get slightly different values according as we suppose the electricity to be distributed over the surface of a conducting sphere of radius a, or rigidly distributed over the

ORIGIN OF THE MASS OF THE CORPUSCLE.

surface of a non-conducting sphere of the same radius, or uniformly distributed throughout the volume of such a sphere. In calculating these differences we have to suppose the charge on the sphere divided up into smaller parts and that each of these small parts obeys the ordinary laws of electrostatics. If we suppose that the charge on the corpuscles is the unit of negative electricity, it is not permissible to assume that smaller portions will obey the ordinary laws of electrostatic attraction.

Perhaps the simplest assumption we can make is that the energy is the same as that outside a sphere of radius a moving with the velocity V and with a charge e at its centre. I have calculated the value of a on this supposition; the results are given in the following Table. The first column of the Table contains the velocity of the corpuscles, which were the object of Kaufmann's experiments; the second column, the values found by Kaufmann for the ratio of the mass of corpuscles moving with this velocity to the mass of a slowly moving corpuscle, and the third column the value of a calculated on the preceding hypothesis.

Velocity of Corpuscle.	Ratio of Mass to that of a Slow Corpuscle.	a
$2{\cdot}85 \times 10^{10}$ cm/sec.	3·09	3·1
$2{\cdot}72 \times 10^{10}$ cm/sec.	2·43	2·42
$2{\cdot}59 \times 10^{10}$ cm/sec.	2·04	2·0
$2{\cdot}48 \times 10^{10}$ cm/sec.	1·83	1·66
$2{\cdot}36 \times 10^{10}$ cm/sec.	1·65	1·5

You will notice that the second and third columns are almost identical; the second column, however, expresses the increase of the whole mass; the third column, the increase of the electrical mass. We see that these are practically equal to each other, hence we conclude that the whole of the mass of the corpuscle is electrical. This electrical mass has its origin in the region round the corpuscle and is not resident in the corpuscle itself; hence, from our point of view, each corpuscle may be said to extend throughout the whole universe, a result which is interesting in connection with the dogma that two bodies cannot occupy the same space.

From the result that the whole of the mass is electrical we are able to deduce the size of the corpuscle, for if m is the mass,

$$m = \frac{2}{3} \frac{e^2}{a}.$$

Now we have seen that $e/m = 1\cdot7 \times 10^7$, and that in electromagnetic measure $e = 10^{20}$. Substituting these values we find that a the radius of the corpuscle $= 10^{-13}$ cm. The radius of the atom is usually taken as about 10^{-8} cm., hence the radius of a corpuscle is only about the one-hundred-thousandth part of the radius of the atom. The potential energy due to the charge is $\frac{1}{2} \frac{e^2 V^2}{a}$, if V is the velocity of light; this potential energy is about the same in amount as the kinetic energy possessed by an α particle moving with a velocity about one-fiftieth that of light.

EVIDENCE OF THE EXISTENCE OF CORPUSCLES AFFORDED BY THE ZEEMAN EFFECT.

The existence of corpuscles is confirmed in a very striking way by the effect produced by a magnetic field on the lines of the spectrum and known as the Zeeman effect. Zeeman found that when the luminous body giving out the spectrum is placed in a strong magnetic field, many of the lines which are single before the application of the field are resolved into three or more components. The simplest case is when a line originally single is resolved into three components, the luminous body being looked at in a direction at right angles to the lines of magnetic force; the middle line of the three occupies its old position, and the side lines are separated from it by an amount proportional to the magnetic force. All the lines

ORIGIN OF THE MASS OF THE CORPUSCLE.

are plane polarised, the plane of polarisation of the middle line being at right angles to that of the side lines. If the same line is looked at in the direction of the magnetic force, the middle line is absent and the two side lines are circularly polarised in opposite senses.

The theory of this simple case, which was first given by Lorentz, is as follows: Let us assume that the vibrating system giving out the line is a charged body, and that it is vibrating under the action of a force whose magnitude is directly proportional to the distance of the vibrating body from a fixed point, and whose direction always passes through the point. Suppose that O is the fixed point and P the electrified body, and let us suppose that the latter is describing a circular orbit round O; let m be the mass of

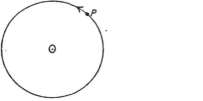

FIG. 14.

the body, $\mu.OP$ the force acting upon it; then the radial acceleration towards O is equal to v^2/OP, v being the velocity of the body. But the product of the mass and the radial acceleration is equal to the radial force $\mu.OP$, hence

$$\frac{m\,v^2}{OP} = \mu.OP$$

If ω is the angular velocity, $v = \omega.OP$, hence

$$\omega^2 = \frac{\mu}{m} \quad \text{or} \quad \omega = \sqrt{\frac{\mu}{m}}$$

The time of vibration is the time OP takes to make a complete revolution or $2\pi/\omega$; thus ω, which is called the frequency of the vibration, is proportional to the number of vibrations per second. In this case the frequency of vibration will evidently be the same whether P goes round O in the direction of the hands of a watch or in the opposite direction. Let now a magnetic force at right angles to the plane of the paper and downwards act upon the charged body. As we have had occasion to remark before, when a charged body moves in a magnetic field it is acted upon by a force which is at right angles to its direction of motion and also to the magnetic force, and equal to $Hev \sin \theta$ where H is the magnetic force, e the charge on the body, v its velocity, and θ the angle between the directions of H and v.

Let now the charged particle be describing a circle in the direction indicated by the arrow round O, the magnetic force being at right angles to the plane of the paper and downwards. The force due to the magnetic field will be radial and in this case directed inwards, and equal to Hev; hence, in addition to the radial force $\mu.OP$, we have the force Hev; equating the product of the mass and the radial acceleration to the radial force we have

$$m \frac{v^2}{OP} = \mu.OP + He.v \qquad (1)$$

and since $v = \omega \times OP$

$$\omega^2 = \frac{\mu}{m} + \frac{He\,\omega}{m} \text{ or } \omega = \frac{1}{2}\frac{He}{m} + \sqrt{\frac{\mu}{m} + \frac{H^2 e^2}{4m}}$$

thus ω is greater than before, and if μ/m is large compared with He/m and equal to ω^2_0 we have

$$\omega = \omega_0 + \frac{1}{2}\frac{He}{m}$$

approximately; ω_0 is the frequency without the magnetic field, thus the change in the frequency is $\frac{1}{2}\frac{He}{m}$, and in this case it is an increase.

Suppose, however, that P were describing the circle in the opposite direction, then, since the direction of motion is reversed the force produced by the magnetic field will be reversed and the force will now be outwards instead of inwards; thus, instead of equation (1) we have

$$\frac{m\,v^2}{} = \mu\,OP - Hev.$$

ORIGIN OF THE MASS OF THE CORPUSCLE. 37

and this treated in the same way as equation (1) leads to the result

$$\omega = \omega_0 - \frac{1}{2}\frac{He}{m}$$

Thus the frequency of vibrations in this direction is *diminished* by an amount equal to that by which the frequency in the opposite direction is increased. Thus the charged body will go round faster in one direction than

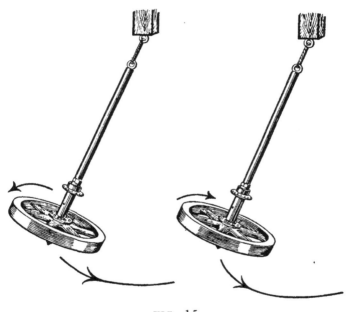

FIG. 15.

in the opposite. I have here an experiment to illustrate a similar effect in a mechanical system. A conical pendulum has for the bob a fly wheel which can be caused to rotate about its axis of rotation. The rotating fly wheel causes a force to act on the bob of the pendulum; this force is at right angles to the direction of motion of the bob, and is proportional to its velocity. It is thus analogous to the force acting on the charged particle due to the magnetic field. The radial force on the electrified particle is represented by the component of gravity at right angles to the axis of the pendulum. I set this pendulum swinging as a conical pendulum with the fly wheel not in rotation. As you would naturally suppose, it goes round just as fast in the direction of the hands of a watch as in the opposite direction. I now set the fly wheel in rapid rotation and repeat the experiment. You see that now the pendulum goes round distinctly more rapidly in one direction than in the opposite, and the direction in which the rotation is most rapid is that in which the rotation of the pendulum is in the same direction as that of the fly wheel.

We see from these considerations that a corpuscle which, when free from magnetic force, would vibrate with the same frequency in whatever direction it might be displaced will no longer do so when placed in a magnetic field. If the corpuscle is displaced so as to move along the lines of magnetic force, the force on the corpuscle due to the magnetic field will vanish, since it is proportional to the sine of the angle between the magnetic force and the direction of motion of the particle; and in this case the frequency will be the same as without the field. When, however, the corpuscle vibrates in the plane at right angles to the lines of magnetic force the frequency will be $\omega + \frac{1}{2}\frac{He}{m}$ if it goes round in one direction, and $\omega - \frac{1}{2}\frac{He}{m}$ if it goes round in the other. Thus in the magnetic field the corpuscles will vibrate with the three frequencies ω, $\omega + \frac{1}{2}\frac{He}{m}$, $\omega - \frac{1}{2}\frac{He}{m}$; one of these being the same as when it was undisturbed. Thus, in the spectroscope there will be three lines instead of one, the middle line being in the undisturbed position. If, however, we look at the corpuscle in the direction of the magnetic force, since the vibrations corresponding to the undisturbed position of the lines are those in which the vibrations are along the lines of magnetic force, and since a vibrating electrified particle does not send out any light along the line of its vibration, no light will come from the corpuscle to an eye

ORIGIN OF THE MASS OF THE CORPUSCLE.

situated along a line of magnetic force passing through the corpuscle, so that in this case the central line will be absent, while the two side lines which correspond to circular orbits described by the corpuscle in opposite directions will give rise to circularly polarised light. By finding the sense of rotation of the light in the line whose frequency is greater than the undisturbed light, it has been shown that the light is due to a negatively electrified body. By measuring the displacement of the lines we can determine the change in frequency, i.e., $\frac{1}{2}\frac{TT_0}{m}$, so that if H is known, e/m can be determined. In this way Zeeman has found the value of e/m to be of the order 10^7, the same as that deduced by the direct methods previously described. The values of e/m got in this way are not the same for all lines of the spectra, but when the lines are divided up into series, as in Paschen and Runge's method, the different lines in the same series all give the same value of e/m.

The displacement of the lines produced by the magnetic field is proportional to e/m, and thus for light due to the oscillations of a corpuscle the displacement will be more than a thousand times greater than that due to the vibration of any positive ion with which we are acquainted. It requires very delicate apparatus to detect the displacement when e/m is 10^7: a displacement one-thousandth part of this would be quite inappreciable by any means at present at our disposal, hence we may conclude that the light in any lines which show the Zeeman effect (and in line spectra as distinct from band spectra, all lines do show this effect to some extent) is due to the vibrations of corpuscles and not of atoms.

The Zeeman effect is so important a method of finding out something about the structure of the atom and the nature of the vibrating systems in a luminous gas, that it is desirable to consider a little more in detail the nature of the conclusions to be drawn from this effect. In the first place it is only a special type of vibration that will show the Zeeman effect. The simple case we considered was when the corpuscle was attracted to O (Fig. 14) by a force proportional to OP; this force is the same in all directions, so that if the corpuscle is displaced from O and then let go it will vibrate in the same period in whatever direction it may be displaced: such a corpuscle shows the Zeeman effect. If, however, the force on P were different in different directions so that the times of vibration of the corpuscle depended on the direction in which it was displaced, then the vibrations would not have shown this effect. The influence of the magnetic force would have been of a lower order altogether than in the preceding case. A single particle placed in a field of force of the most general character might vibrate with three different periods and thus give out a spectrum containing three lines, but if such a particle were placed in a magnetic field these lines would not show the Zeeman effect; all that the magnetic force

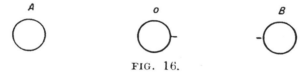

FIG. 16.

could do would be to slightly alter the periods by an amount infinitesimal in comparison with that observed in the Zeeman effect. There could be no resolution of the lines into triplets; it is only in the special case when the periods all become the same that the Zeeman effect occurs. We can easily imagine cases in which some lines might show the Zeeman effect, while others would not do so. Take the case of two corpuscles A and B attracted to a point O (Fig. 16) and repelling each other, they will settle into a position of equilibrium when the repulsion between them balances the attraction exerted by O. In the most general case there would be six different frequencies of vibration (each corpuscle contributing three) and none of these would show the Zeeman effect. In the special case where the force exerted by O is the same in all directions, three of these frequencies coincide, two others vanish, and there is one remaining isolated. The spectrum is reduced to two

ORIGIN OF THE MASS OF THE CORPUSCLE. 41

lines: one of these (that corresponding to the coalescence of the three lines) would show the normal Zeeman effect while the other would not show it at all. With more complicated systems we might have several lines showing the Zeeman effect accompanied by others which do not show it. When more lines than one show the Zeeman effect, the magnitude of the effect may differ from line to line. Thus, take the case of four corpuscles mutually repelling each other and attracted towards a point O. In the most general case this system would have twelve different frequencies, three for each corpuscle, and as long as these remained different none of them would show the Zeeman effect. If, however, the force exerted by O is the same in all directions, two sets of three of these frequencies become equal, three frequencies vanish, two others coincide, and one remains isolated; the twelve frequencies are now reduced to four, the two lines corresponding to the sets of three frequencies which had coalesced will both show the Zeeman effect, but not to the same extent, the alteration in frequency for one line being the normal amount $\frac{1}{2}\frac{He}{m}$ while for the other line it is only half that amount. The other lines do not show the Zeeman effect. The reader who is interested in this subject is referred for other instances of systems illustrating this effect to a paper by the writer in the Proceedings of the Cambridge Philosophical Society, vol. xiii., p. 39.

It is remarkable that, as far as our knowledge extends, all the lines in a line spectrum show the Zeeman effect. This might arise from the vibrating systems being single corpuscles, only influenced slightly, if at all, by neighbouring corpuscles, or it might arise from the vibrations of more complicated systems, provided the radiation corresponding to frequencies which on the theory would not show the Zeeman effect, has great difficulty in leaving the vibrating system. We have an example of the second condition in the case of two corpuscles shown in Fig. 16; the vibration which does not show the Zeeman effect is the one when the middle point of A and B remains at rest and A and B are approaching O or receding from it with equal velocities; thus the charged corpuscles are moving with equal velocities in opposite directions and their effects, at a distance from O large compared with OA and OB will neutralise each other. On the other hand, the vibrations which show the Zeeman effect are those in which A and B are moving in the same direction, so that the effects due to one will supplement those due to the other, and thus the intensity of the radiation from this vibration will greatly exceed that from the other; thus this vibration might give rise to visible radiation while the other did not. The vibration of a system of corpuscles which produces the greatest effect at a distance, is the one where all the corpuscles move with the same speed and in the same direction; it can be easily shown that for this case the effect of a magnetic field is to increase or diminish all the frequencies by the normal amount $\frac{1}{2}\frac{H}{m}$.

A case in which the Zeeman effect might be abnormally large is the following:—Suppose we have two corpuscles A and B moving round the circumference of a circle with constant angular velocity ω, always keeping at opposite ends of a diameter, then the frequency of the optical or magnetic effect produced by this system is not ω but 2ω, for each particle has only to go half way round the circumference to make the state of the system recur. If now we place the system in a magnetic field where the magnetic force is perpendicular to the circle the angular velocity ω will become $\omega + \frac{1}{2}\frac{H}{m}$ and the frequency of the system $2\omega + \frac{H}{m}$, thus the change in the frequency is $\frac{H}{m}$, which is twice the normal effect.

CHAPTER III.

PROPERTIES OF A CORPUSCLE.

HAVING demonstrated the existence of corpuscles, it will be convenient for purposes of reference to summarise their properties.

MAGNETIC FORCE DUE TO CORPUSCLES.

A moving corpuscle produces around it a magnetic field. If the corpuscle is moving in a straight line with a uniform velocity v, which is small compared with the velocity of

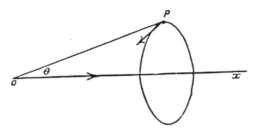

FIG. 17.

light, it produces a magnetic field in which the lines of magnetic force are circles having the line along which the corpuscle is moving for their axis; the magnitude of the force at a point P is equal to $\frac{ev}{OP^2} \sin \theta$, where e is the charge on the moving particle O, and θ the angle between OP and OX—the line along which the corpuscle is moving. The direction of the force at P (Fig. 17) is at right angles to the plane POX and downwards from the plane of the paper if the *negatively* charged particle is moving in the direction OX. The magnetic force thus vanishes along the line of motion of the particle and is greatest in the plane through O at right angles to the direction of motion; the distribution of force is symmetrical with respect to this plane.

If the velocity of the uniformly moving particle is so great that it is comparable with c the velocity of light, the intensity of the magnetic force at P is represented by the more complicated expression

$$\left(1 - \frac{v^2}{c^2}\right) \frac{e\, v\, \sin\theta}{r^2 \left(1 - \frac{v^2}{c^2} \sin^2\theta\right)^{\frac{3}{2}}}$$

The direction of the force is the same as before. The effect of the greater velocity is to make the magnetic force relatively weaker in the parts of the field near OX and stronger in those near the equatorial plane, until when the speed of the corpuscle is equal to that of light the magnetic force is zero everywhere except in the equatorial plane, where it is infinite.

Electric Field round the Moving Corpuscle.

The direction of the electric force at P is along OP, and whatever be the speed at which the corpuscle is moving, the electric force E and the magnetic force H are connected by the relation

$$c^2 H = v E \sin\theta;$$

thus when the corpuscle is moving slowly

$$E = \frac{e\, c^2}{r^2}$$

the same value as when the particle is at rest (remembering that e is measured in electro-magnetic units).

When the corpuscle is moving more rapidly we have

$$E = (c^2 - v^2) \frac{e}{r^2 \left(1 - \frac{v^2}{c^2} \sin^2\theta\right)^{\frac{3}{2}}}$$

and in this case the electric force is no longer uniformly distributed, but is more intense towards the equatorial regions than in the polar regions near OX. When the

PROPERTIES OF A CORPUSCLE.

corpuscle moves with the velocity of light all the lines of electric force are in the plane through O at right angles to OX.

When the corpuscle is moving uniformly the lines of force are carried along as if they were rigidly attached to it, but when the velocity of the corpuscle changes this is no longer the case, and some very interesting phenomena occur. We can illustrate this by considering what happens if a cor-

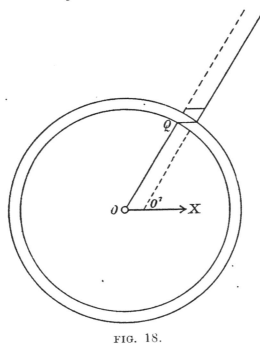

FIG. 18.

puscle which has been moving uniformly is suddenly stopped. Let us take the case when the velocity with which the particle is moving before it is stopped is small compared with the velocity of light; then before the stoppage the lines of force were uniformly distributed and were moving forward with the velocity v. When the corpuscle is stopped, the ends of the lines of force on the corpuscle will be stopped also; but fixing one end will not at once stop the whole of the line of force, for the impulse which stops the tube travels along the line of force with the velocity of light, and thus takes a finite time to reach the outlying parts of the tube. Hence when a time t has elapsed after the stoppage, it is only the parts of the lines of force which are inside a sphere whose radius is ct which have been stopped. The lines of force outside this sphere will be in the same position as if the corpuscle had not been stopped, *i.e.*, they will pass through O', the position the corpuscle would have occupied at the time t if the stoppage had not taken place. Thus the line of force which, when the corpuscle was stopped was in the position OQ will, at the time t be distorted in the way shown in Fig 18. Inside the sphere of radius ct the line of force will be at rest along OQ; outside the sphere it will be moving forward with the velocity v, and will pass through O', the point O would have reached at the time t if it had not been stopped. Since the line of force remains intact it must be bent round at the surface of the sphere so that the portion inside the sphere may be in connection with that outside. Since the lines of force along the surface are tangential there will be, over the surface of the sphere, a tangential electric force. This tangential force will be on the surface of a sphere of radius ct and will travel outwards with the velocity of light. If the stoppage of the sphere took a short time π, then the tangential part of the lines of force will be in the spherical shell between the spheres whose radii are ct and $c(t-\pi)$, t being the time which has elapsed since the stoppage began, and $t-\pi$ since it was completed. This shell of thickness $c\pi$, filled with tangential lines of electric force, travels outwards with the velocity of light. The electric force in the shell is very large compared with the force in the same region before the shell is stopped. We can prove that the magnitude of the force at a point P in the shell is equal $\dfrac{c \, e \, v \, \sin \theta}{OP \cdot \delta}$, where δ is the thickness of the shell, and θ the angle POX. Before the corpuscle was stopped the force was $\dfrac{c^2 e}{OP^2}$, thus the ratio of the force after the stoppage to the force before is equal to $\dfrac{v \, OP}{c \, \delta} \sin \theta$. As δ

PROPERTIES OF A CORPUSCLE.

is very small compared with OP, this ratio is very large; thus the stoppage of the corpuscle causes a thin shell of intense electric force to travel outwards with the velocity of light. These pulses of intense electric force constitute, I think, Rontgen rays, which are produced when cathode rays are suddenly stopped by striking against a solid obstacle. The electric force in the pulse is accompanied by a magnetic force equal in magnitude to $\frac{v \, su}{OP \cdot o}$ – and at right angles to the plane POX. The energy in the pulse due to this distribution of magnetic and electric force is equal to $\frac{2}{3} \frac{e^2 \, v^2}{o}$; it is thus greater when the thickness of the pulse is small than when it is large. The thickness of the pulse is, however, proportional to the abruptness with which the corpuscle is stopped; and as the energy in the pulse is radiated away it follows that the more abruptly the corpuscles are stopped the greater the amount of energy radiated away as Rontgen rays. If the corpuscle is stopped so abruptly that the thickness of the pulse is reduced to the diameter of the corpuscle the whole of the energy in the magnetic field round the corpuscle is radiated away. If the corpuscle is stopped more slowly only a fraction of this energy escapes as Rontgen rays.

Inside the shell, *i.e.*, in the space bounded on the outside by the sphere of radius OP ($=ct$), there is no magnetic force, while outside the sphere whose radius is OP the magnet force is the same as it would be if the particle had not been stopped, *i.e.*, at the point Q it is equal to $\frac{\ldots}{O'O^2} \sin \phi$, where O' is where O would have been if the corpuscle had gone on moving uniformly, and ϕ is the angle $QO'X$. The pulse in its outward passage wipes out, as it were, the magnetic force from each place as it passes over it.

We have seen that when the corpuscle is stopped there is a pulse of strong electric and magnetic force produced which carries energy away. It is not necessary that the

48 THE CORPUSCULAR THEORY OF MATTER.

corpuscle should be reduced to rest for this pulse to be formed; any change in the velocity will produce a similar pulse, though the forces in the pulse will not be so intense as when the stoppage is complete. Since any change in the velocity produces this tangential electric field, such a field is a necessary accompaniment of a corpuscle whose motion is accelerated, and we can show that if when at O the particle has an acceleration f along OX, then after a time t has elapsed there will be at a point P distant ct from O a tangential electric force equal to $\dfrac{e\, f \sin \theta}{OP}$ and a magnetic force at right angles both to OP and the electric force, equal to $\dfrac{e\, f \sin \theta}{C \cdot OP}$. The rate at which energy is being radiated from the corpuscle has been shown by Larmor to be equal to $\tfrac{2}{3} \dfrac{e^2 f^2}{C}$; thus a corpuscle whose velocity is changing loses energy by radiation.

CHAPTER IV.

CORPUSCULAR THEORY OF METALLIC CONDUCTION.

WE now proceed to apply these properties of corpuscles to the explanation of some physical phenomena; the first case we shall take is that of conduction of electricity by metals.

On the corpuscular theory of electric conduction through metals the electric current is carried by the drifting of negatively electrified corpuscles against the current. Since the corpuscles and not the atoms of the metal carry the current, the passage of the current through the metal does not imply the existence of any transport of these atoms along the current; this transport has often been looked for but never detected. We shall consider two methods by which this transport might be brought about.

In the *first method* we suppose that all the corpuscles which take part in the conduction of electricity have got into what may be called temperature equilibrium with their surroundings, *i.e.*, that they have made so many collisions that their mean kinetic energy has become equal to that of a molecule of a gas at the temperature of the metal. This implies that the corpuscles are free not merely at the instant the current is passing but that at this time they have already been free for a time sufficiently long to allow them to have made enough collisions to have got into temperature equilibrium with the metal in which they are moving. The corpuscles we consider are thus those whose freedom is of long duration. On this view the drift of the corpuscles which forms the current is brought about by the direct action of the electric field on the free corpuscles.

Second Method.—It is easy to see, however, that a

current could be carried through the metal by corpuscles which went straight out of one atom and lodged at their first impact in another; such corpuscles would not be free in the sense in which the word was previously used and would have no opportunities of getting into temperature equilibrium with their surroundings. To see how conduction could be brought about by such corpuscles, we notice that the liberation of corpuscles from the atoms must be brought about by some process which depends upon the proximity of the metallic atoms. We see this because the ratio of the conductivity of a metal in a state of vapour to the conductivity of the same metal when in the solid state is exceedingly small compared with the ratio of the densities in the two states. Some interesting experiments on this point have been made by Strutt, who found that when mercury was heated in a vessel to a red heat, so that the pressure and density must have been exceedingly large, the conductivity of the vapour was only about one-ten millionth of the conductivity of solid mercury. If, however, corpuscles readily leave one atom and pass into another when the atoms of the metal are closely packed together, we can see how the electricity could pass without any accumulation of free corpuscles. For, to fix our ideas, imagine that the atoms of the metal act on each other as if each atom were an electric doublet, *i.e.*, as if it had positive electricity on one side and negative on another. A collection of such atoms if pressed close together would exert considerable force on each other, and the force exerted by an atom A on another B, might cause a corpuscle to be torn out of B. If this got free and knocked about for a considerable time it would form one of the class of corpuscles previously considered, but even if it went straight from B into A it might still help to carry the current. If the atoms were arranged without any order, then, though there might be interchange of corpuscles between neighbouring atoms, there would be no flux of corpuscles in one direction rather than another, and therefore no current. Suppose, however, that the atoms get polarised under the action of an electric force, which force is, say, horizontal and from left to right, then the atoms will have a tendency to arrange themselves so that the negative ends are to the left, the positive ones to the right. Consider two neighbouring atoms A and B (Fig. 19): if a corpuscle is dragged out of A into B it will start from the negative end of A and go to the positive end of B; there will thus be more corpuscles going from right to left than in any other direction; this will give rise to a current from left to right, *i.e.*, in the direction of the electric force.

We shall develop the consequences of each of these theories so as to get material by which they can be tested.

A piece of metal on the first of these theories contains a large number of free corpuscles disposed through its volume. These corpuscles can move freely between the atoms of the metal just as the molecules of air move freely about in the

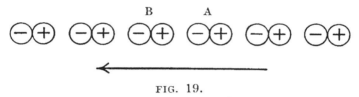

FIG. 19.

interstices of a porous body. The corpuscles come into collision with the atoms of the metal and with each other and at these impacts suffer changes in velocity and momentum; in fact, these collisions play just the same part as the collisions between molecules do in the kinetic theory of gases. In that theory it is shown that the result of such collisions is to produce a steady state in which the mean kinetic energy of a molecule depends only upon the absolute temperature: it is independent of the pressure or the nature of the gas, thus it is the same for hydrogen as for air. We may regard the corpuscles as being a very light gas, so that the mean kinetic energy of the corpuscles will only depend upon the temperature and will be the same as the mean kinetic energy of a molecule of hydrogen at that temperature. As, however, the mass of a corpuscle is only about 1/1700 of that of an atom of hydrogen, and therefore only about 1/3400 of that of a molecule of hydrogen, the mean value of the square of the velocity of a corpuscle must be 3400 times that of the same quantity for the molecule of hydrogen at the same temperature. Thus the average velocity of the corpuscle must be about 58 times that of a molecule of hydrogen at the temperature of the metal in which the molecules are situated. At 0° C. the mean velocity of the hydrogen molecule is about $1·7 \times 10^5$ cm/sec, hence the average velocity of the corpuscles in a metal at this temperature is about 10^7 cm/sec, or approximately 60 miles per sec. Though these corpuscles are charged, yet since as many are moving in one direction as in the opposite, there will be on the average no flow of electricity in the metal. The case is, however, altered when an electric force acts throughout the metal. Although the change produced in the velocity of the corpuscles by this force is, in general, very small compared with the average velocity of translation of the corpuscles, yet it is in the same direction for all of them, and produces a kind of wind causing the corpuscles to flow in the opposite direction to the electric force (since the charge on the corpuscle is negative), the velocity of the wind being the velocity imparted to the corpuscles by the electric force. If u is this velocity and n the number of corpuscles per unit volume of the metal, the number of corpuscles which in one second cross a unit area drawn at right angles to the electric force is $n\,u$, and if e is the charge on a corpuscle, the quantity of electricity carried through this area per second is $n\,u\,e$; this quantity is the intensity of electric current in the metal; if we denote it by i, we have the equation $i = n\,u\,e$. We now proceed to find u in terms of X the electric force in the metal. While the corpuscle is moving in a free path in the interval between two collisions, the electric force acts upon it and tends to make it move in the opposite direction to itself. When, however, a collision occurs, the shock is so violent that the corpuscle moves off in much the same way, and with much the same velocity, as if it had not been under the electric field. Thus the effect of the electric field is, so to speak, undone at each collision; after the collision the electric force has to begin again, and the velocity communicated by the electric field to the corpuscle will be that which it gives to it during its free path. Jeans has shown that there is a slight persistence of an effect produced on a molecule after an encounter with another molecule, that each collision does not, as it were, entirely wipe out all the effects of the previous history of the molecule. To calculate the amount of this persistence we have to know the nature of the effect we call a collision; in our case the effect is not of importance. If m is the mass of the corpuscle, the velocity the corpuscle owes to the action of the electric force increases uniformly from zero at the beginning of the free path to $X\frac{e}{m}t$ at the end, t being the time between two collisions; hence the mean velocity due to the force is $\frac{1}{2}X\frac{e}{m}t$, and this is the velocity given to the particles by the electric force. If we care to take into account the persistence of the impression produced by the electric force we can do so by introducing a factor β into the expression and saying that the average velocity u due to the electric field is $\frac{1}{2}\beta\frac{Xe}{m}t$. Unless, however, we have a knowledge of the nature of the collision between a corpuscle and the atom, all that we can determine about β is that it is a quantity somewhat greater than unity. Since $u = \frac{1}{2}\beta\frac{Xe}{m}t$, and $i = nue$, we have

$$i = \frac{1}{2}\beta\frac{Xe^2 t}{m}.$$

Now unless the electric force is enormously large the change in the velocity of the corpuscle due to the electric force will be quite insignificant in comparison with v—the average velocity of translation of the corpuscle. We may therefore put $t = \lambda/v$, where λ is the mean free path of a corpuscle, hence

$$i = \frac{1}{2}\beta n\frac{Xe^2}{m}\frac{\lambda}{v} = \frac{1}{2}\beta n\frac{Xe^2 \lambda v}{m v^2}.$$

Now $m v^2$ is twice the average kinetic energy of a

54 THE CORPUSCULAR THEORY OF MATTER.

corpuscle, and therefore twice the kinetic energy of a molecule of hydrogen at the same temperature; $m v^2$ is thus equal to $2 a \theta$ where θ is the absolute temperature and $2a = 7\cdot2 \times 10^{-14}/273$.

From the relation—

$$i = \frac{1}{2} \beta n \frac{X e^2 \lambda v}{m v^2} = \frac{1}{4} \frac{\beta}{a \theta} n e^2 \lambda v X$$

we see that the specific conductivity of the metal is equal to $\beta n e^2 \lambda v / 4 a \theta$; thus the specific conductivity on this theory is independent of the electric force X, so that Ohm's law is true.

If the electric force were so large that the velocity generated in a corpuscle during its free path were large compared with the average velocity of a corpuscle, the relation between current and electric force would take a different form. In this case the velocity of the particle is generated by the field, so that if w is this velocity then $\frac{1}{2} m w^2 = X e \lambda$, or $w = \sqrt{\frac{e X \lambda}{m}}$; the average velocity is one-half of this, i.e., $\sqrt{\frac{X e \lambda}{2m}}$, and the current $i = n e \sqrt{\frac{X e \lambda}{2m}}$. Thus in this case the current, instead of being proportional to the electric force, would be proportional to the square root of it, so that Ohm's law would no longer hold. This state of affairs would, however, only occur when the electric force was exceedingly large, too large to be realised by any means we have at present at our command. For it requires $X e \lambda$ to be large compared with the mean kinetic energy of a corpuscle, which at 0° C. is equal to $3\cdot6 \times 10^{-14}$. Now e is 10^{-20}, thus $X \lambda$ must be large compared with $3\cdot6 \times 10^6$. We do not know the free path of a corpuscle in a metal, but as the free path in air whose density at atmospheric pressure is only ·0015 is only 10^{-5} cms., the free path in a metal can hardly be greater than 10^{-7} cms. Thus the value of X necessary to give to

THEORY OF METALLIC CONDUCTION. 55

the corpuscle an amount of kinetic energy large compared with that it possesses in virtue of the temperature of the metal, must be of the order 10^{14}, *i.e.*, a million volts per centimetre. We have no experimental evidence as to how a conductor would behave under forces of this magnitude.

If we assume that λ is of the order 10^{-7} we can get an estimate of n—the number of corpuscles in a cubic centimetre of the metal. Let us take for example silver, whose specific conductivity is 1/1600 at 0° C.; we have, using the expression we have found for the conductivity—

$$\frac{1}{1600} = \frac{\beta}{4} \frac{n \, e^2 \, \lambda \, v}{a \, \theta};$$

if we put $e = 10^{-20}$, $\lambda = 10^{-7}$, $v = 10^7$, $\beta = 1$, $2 a \theta = 7{\cdot}2 \times 10^{-14}$ we find $n = 9 \times 10^{23}$.

Now, in a cubic centimetre of silver there are about $1{\cdot}6 \times 10^{23}$ atoms of silver, and thus from this very rough estimate we conclude that even in a good conductor like silver the number of corpuscles is a quantity comparable with the number of atoms.

If the carriers instead of being corpuscles were bodies with a greater mass the number of carriers would be greater than that just found. For we see from the preceding formula that if the carriers are in temperature equilibrium with the metal $n \lambda v$ must be constant if the conductivity is given. Hence if the mass of the carriers were much greater than that of a corpuscle and therefore v and λ much smaller, n would have to be much larger, that is, the number of carriers in silver would have to be much greater than the number of atoms of silver, a result which shows that the mass of a carrier cannot be comparable with that of an atom.

COMPARISON OF THE THERMAL WITH THE ELECTRICAL CONDUCTIVITY.

If one part of the metal is at a higher temperature than another, the average kinetic energy of the corpuscles in the hot parts will be greater than that in the cold. In consequence of the collisions which they make with the atoms of the metal, resulting in alterations in the energy, the corpuscles will carry heat from the hot to the cold parts of the metal; thus a part at least of the conduction of heat through the metal will be due to the corpuscles. If we assume that the whole of the conduction arises in this way, we can find an expression for the thermal conductivity in terms of the quantities which express the electrical conductivity. It is proved in treatises on the kinetic theory of gases that k the thermal conductivity of a gas is given by the expression—

$$k = \tfrac{1}{3} n \lambda v a$$

(see Jean's "Kinetic Theory of Gases," p. 259). Here k is measured in mechanical units, and the effect of persistence of the velocities after the collisions has been neglected. Hence to compare k with c the electrical conductivity we must in the expression for the latter quantity put $\beta = 1$; doing this, we obtain—

$$c = \frac{n \lambda v e^2}{4 a \theta},$$

hence—

$$k/c = \frac{4}{3} \cdot \frac{a^2 \theta}{e^2}.$$

Thus neither n nor λ, the quantities which vary from metal to metal, appears in the expression for c/k, so that the theory of corpuscular conduction leads to the conclusion that the ratio of the electrical to the thermal conductivity should be the same for all metals and should vary inversely as the absolute temperature of the metals.

We can calculate the numerical value of the ratio of the two conductivities on the preceding theory as follows: If p is the pressure of a gas in which there are n molecules per cubic centimetre, θ the absolute temperature, then—

$$p = \tfrac{2}{3} a \theta . n;$$

hence—

$$\frac{a \theta}{e} = \frac{3 p}{2 n e}.$$

Now e is the charge on an atom of hydrogen, and if n is the number of hydrogen molecules in a cubic centimetre of gas at a pressure of one atmosphere (*i.e.*, 10^6 dynes), and

THEORY OF METALLIC CONDUCTION.

at 0° C., we have, since one electromagnetic unit of electricity liberates 1·2 cubic centimetres of hydrogen at this pressure and temperature—

$$2·4\, ne = 1;$$

hence at 0° C.—

$$\frac{ne}{c} = 3·6 \times 10^6,$$

so that at this temperature—

$$\frac{k}{c} = \frac{4}{3} \frac{a^2 \theta^2}{c} \frac{\cdot}{\cdot} = 6·3 \times 10^{10} \text{ in absolute measure.}$$

The following are the values of k/c for a large number of metals as determined by Jaeger and Diesselhorst in their most valuable paper on this subject :—

Material.	Thermal conductivity. / Electrical conductivity.	Temperature coefficient of this ratio.
	At 18° C.	Per cent.
Copper, commercial	$6·76 \times 10^{10}$	—
Copper (1), pure	$6·65 \times 10^{10}$	0·39
Copper (2), pure	$6·71 \times 10^{10}$	0·39
Silver, pure	$6·86 \times 10^{10}$	0·37
Gold (1) ...	$7·27 \times 10^{10}$	0·36
Gold (2), pure	$7·09 \times 10^{10}$	0·37
Nickel	$6·99 \times 10^{10}$	0·39
Zinc (1) ...	$7·05 \times 10^{10}$	0·38
Zinc (2), pure	$6·72 \times 10^{10}$	0·38
Cadmium, pure ...	$7·06 \times 10^{10}$	0·37
Lead, pure	$7·15 \times 10^{10}$	0·40
Tin, pure	$7·35 \times 10^{10}$	0·34
Aluminium	$6·36 \times 10^{10}$	0·43
Platinum (1)	$7·76 \times 10^{10}$	—
Platinum (2), pure	$7·53 \times 10^{10}$	0·46
Palladium	$7·54 \times 10^{10}$	0·46
Iron (1) ...	$8·02 \times 10^{10}$	0·43
Iron (2) ...	$8·38 \times 10^{10}$	0·44
Steel	$9·03 \times 10^{10}$	0·35
Bismuth ...	$9·64 \times 10^{10}$	0·15
Constantan (60 Cu 40 Ni)	$11·06 \times 10^{10}$	0·23
Manganine (84 Cu 4 Ni 12 Mn)	$9·14 \times 10^{10}$	0·27

58   THE CORPUSCULAR THEORY OF MATTER.

It will be seen that the observed values of the ratios of the thermal and electrical conductivities of many metals agree closely with the result deduced from theory, while others show considerable deviations. Again, the temperature coefficient of this ratio is for many metals in agreement with the theory. On the theory the ratio is proportional to the absolute temperature; this gives a temperature coefficient of ·366 per cent., and we see that for many metals the temperature coefficient is of this order.

In the case of alloys the ratio of the thermal to the

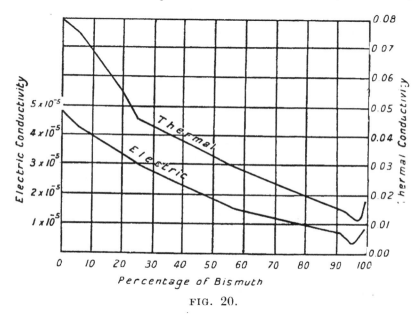

FIG. 20.

electrical conductivity is not nearly so constant as it is for pure metals. Even with alloys, however, any considerable variation in the electrical conductivity is accompanied by a corresponding variation in the thermal conductivity; this is illustrated by the curves given in Fig. 20, taken from a paper by Schulze (*Ann. der Phy.*, ix., p. 584), and which represent the variations in both the electrical and thermal conductivities of alloys of bismuth and lead with the percentage amount of bismuth in the alloy. It will be noticed that the two curves are approximately parallel and have their minimum ordinate at about the same place. As a rule, although there are some exceptions, the ratio of the thermal to the electrical conductivity is larger for alloys than for pure metals. This and many other properties of conduction of electricity through alloys can be explained by some considerations given by Lord Rayleigh (*Nature*, LIV., p. 154, " Collected Works," vol. iv., p. 232). Lord Rayleigh points out that in the case of a mixture of metals there is, owing to their thermo-electric properties, a source of something which cannot be distinguished by experiments from resistance, which is absent when the metals are pure. To see this, let us suppose that the mixed metals are arranged in thin layers, the adjacent layers being of different metals, and that the current passes through the body at right angles to the faces of the layer. Now when a current of electricity passes across the junction of two metals Peltier showed that the junction was heated if the current passed one way, cooled if it passed the opposite way, and that the rate of heat production or absorption was proportional to the current passing across the junction. Thus, where the current passes through the system of alternate layers of the two metals, one face of each layer will be cooled and the other heated, and thus in the pile of layers differences of temperature proportional to the current will be established. These will set up a thermoelectric force, tending to oppose the current, proportional to the intensity of the current. Such a force would produce exactly the same effect as a resistance. Thus in a mixture of metals there is, in addition to the resistance, a 'false resistance' due to thermo-electric causes which is absent in the case of pure metals. This false resistance being superposed on the other resistance makes the electrical resistance of alloys greater than the value indicated by the preceding theory. This result gives an explanation of the fact that the ratio of the thermal to the electrical resistance is greater for alloys than it is for pure metals.

The experiments of Dewar and Fleming on the effect of very low temperatures on the resistance of pure metals and alloys show that there is a fundamental difference between the resistances of pure metals and mixtures, for while the resistance of pure metals diminishes uniformly as the temperature diminishes and would apparently vanish not far from the absolute zero of temperature, the resistance of alloys gives no indication of disappearing at these very low temperatures, but apparently tends to a finite limit.

The electrical conductivity of a metal is proportional to n the number of free corpuscles per unit volume. Now, since a free corpuscle will continually be getting caught by and attached to an atom, the corpuscles, when the metal is in a steady state, must be in statical equilibrium; the number of fresh corpuscles produced in unit time being equal to the number which disappear by re-combination with the atoms in the same time. We should expect the number of re-combinations in unit time to be proportional to the number of collisions in that time, i.e., to n/τ; where τ is the interval between two collisions; τ is equal to λ/v where λ is the free path and v the velocity of the corpuscle. Hence the number of re-combinations in unit time will be equal to $\gamma \frac{n v}{\lambda}$ where γ represents the proportion between the number of collisions which result in re-combination and the whole number of collisions. If q is the number of corpuscles produced per cubic centimetre per second, we have when there is statical equilibrium—

$$q = \gamma \frac{n v}{\lambda}.$$

Thus c the electrical conductivity of a metal is expressed by the equation—

$$c = \frac{1}{4} \frac{\beta}{\gamma} \frac{q \lambda^2 e^2}{\alpha \theta}.$$

For most pure metals the conductivity is inversely proportional to the absolute temperature θ, hence we conclude that $q \lambda^2$ must be independent of the temperature. Now we should not expect λ to vary more rapidly with the temperature than the distance between two molecules, a

THEORY OF METALLIC CONDUCTION. 61

quantity whose variation with the temperature is of the same order as that of the linear dimensions of the body, and therefore represented by the coefficient of thermal expansion, a very small quantity; thus, since $q \lambda^2$ is independent of the temperature, and λ^2 only varies slowly with the temperature, the variations of q with temperature can only be slight, hence we conclude that the dissociation of the atom which produces the corpuscles cannot to any considerable extent be the effect of temperature.

We should expect to have fewer free corpuscles and therefore smaller conductivity in a salt of the metal than in the metal itself. For in the salt the atoms of the metal are all positively electrified and have already lost corpuscles, which have found a permanent home on the atoms of the electro-negative element. From the positively electrified metal atoms corpuscles will find it difficult to escape, and the rate of production of free corpuscles will be very much lower than in the pure metal, where in addition to positively electrified atoms neutral and negatively electrified atoms of the metal are present.

Lorentz Theory of Radiation.

Radiation of heat may be produced by the impact of corpuscles. When a corpuscle comes into collision with an atom it experiences rapid changes in its velocity, and therefore will, as explained on p. 46, emit pulses of intense electric and magnetic force; the thickness of these pulses will be the distance traversed by light during the time occupied by a collision. Thus, if we consider any atom of the metal, it will be from time to time, as the corpuscles strike against it, the centre of pulses of intense electric and magnetic force. These forces at a point near the atom will vary in a very abrupt manner. A pulse of intense electric force, lasting for a very short time, will pass over the point, then there will be an interval in which the electric force disappears, and again, after the space of time between two collisions, another intense pulse will pass over the point. Now though the electric force jumps about in this abrupt way, we know by the theorem due to Fourier that it can be represented as the sum of a number of terms, each of which is of the form $cos\,(pt + e)$ where t represents the time. Each of these terms represents a harmonic wave of electric force, and by the electro-magnetic theory of light a harmonic wave of electric force is a wave of light or radiant heat. Thus we can represent the irregular, jerky electric field produced by the collision as arising from the superposition of a number of waves of light or radiant heat, and if we can calculate the amplitude of vibration of the disturbance of any period, we can calculate at once the energy in the light of this period emitted by one molecule, and therefore, by summation, by the metal.

Of the whole group of waves which represent the electric field due to the collisions, Lorentz has shown how to calculate the amplitudes of those whose wave length is very large indeed compared with the free path of the corpuscles, and has shown that the energy in the vibrations whose frequency is between q and δq given out per second per unit of area of a plate where thickness is Δ is equal to—

$$\frac{\Delta\, q^2\, dq}{6\, \pi^2\, c}\; 4\,\pi\, e^2\, n\, \lambda\, v\,;$$

c represents the velocity of light, e the charge on the corpuscle, λ the mean free path of a corpuscle, and v its mean velocity of translation. This represents the rate at which the body emits energy. To find the amount of energy of this frequency present in the body when the radiation is in a steady state, we must take into account the absorption of this energy in its course through the body. For imagine a body built up of piles of parallel plates; then if there were no absorption the energy emitted by the most distant portions would reach any point Q, and if the size of the body were infinite the amount of energy per unit volume at Q would be infinite also. If, however, there was strong absorption, so that the radiation was practically all absorbed in the space of one millimetre, then it is evident

THEORY OF METALLIC CONDUCTION. 63

that the portions of the body whose distance from Q is more than one millimetre will not send any energy to Q, and however large the body may be the energy at Q will be finite. When the energy in the body has settled down into a steady state, the energy given out by any portion must be equal to the amount acquired by absorption. This principle enables us to find the amount of energy per unit volume of the body when the radiation is in a steady state. The absorption of these very long waves in a conductor is due to the same cause as the production of heat in the conductor when an electric current passes through it, since these waves are made up of electric and magnetic forces. When an electric force X acts on a conductor and produces an electric current whose intensity is i, the rate at which energy is absorbed per unit volume is Xi, or if σ is the specific resistance of the conductor the rate at which energy is absorbed is equal to X^2/σ, since $\sigma i = X$. We must express this in terms of E the energy per unit volume in the conductor. One half of this energy is due to the electric field, the other half to the magnetic field which accompanies it; the energy per unit volume due to the electric field is $\dfrac{X^2}{8\pi c^2}$, c being the velocity of light through the medium, hence $E = \dfrac{X^2}{4\pi c^2}$, and $X^2 = 4\pi c^2 E$, hence X^2/σ the rate at which energy is absorbed per unit volume is equal to—

$$\frac{4\pi c^2 E}{\sigma},$$

and the rate per unit area of a plate of thickness Δ is—

$$\frac{4\pi c^2 E \Delta}{\sigma}.$$

Now in a steady state the energy emitted is equal to the energy absorbed; the expression for the rate at which energy is emitted is given on p. 62; equating this to the rate at which the energy is absorbed, we have—

$$\frac{4\pi c^2 E \Delta}{\sigma} = \frac{\Delta q^2 dq}{6\pi^2 c} 4\pi e^2 \qquad (1)$$

but (see p. 56)—

$$\frac{1}{\sigma} = \frac{e^2 \lambda n v}{4 a \theta}$$

when θ is the absolute temperature. Substituting this value for $1/\sigma$, equation (1) becomes—

$$c^2 E \frac{(e^2 \lambda n v)}{4 a \theta} = \frac{q^2 d q}{6 \pi^2 c} e^2 n \lambda v. \qquad (2)$$

The quantities n and λ which differentiate one substance from another occur in the same form on both sides of the equation: one side expresses the absorption, the other the radiation, and we see that the ratio of the two is independent of the nature of the substance. Hence this view of radiation would explain Kirchhoff's law that good radiators are also good absorbers. Dividing out the common factors from equation (2), we get—

$$E = \frac{2}{3} \frac{a \theta}{\pi^2 c^3} q^2 d q;$$

or if λ is the wave length of the vibration whose frequency is q we have, since—

$$q = 2 \pi \frac{c}{\lambda},$$

$$E = \frac{16}{3} \frac{\pi a \theta}{\lambda^4} d \lambda,$$

and this is the expression for the amount of energy per unit volume whose wave length is between λ and $d \lambda$ when the absolute temperature is θ. This expression does not involve any constant which depends upon the nature of the body, hence it would be the same at the same temperature for all bodies. The expression for E is of the type $f \cdot (\lambda \theta) \frac{d \lambda}{\lambda^5}$, where $f (\lambda \theta)$ denotes a function of λ and θ. The researches of Wien have shown that it is only a formula of this type which fits in with the values of the radiation observed by him and others in experiments with bodies at different temperatures. The preceding expression is of the type suggested by Lord Rayleigh (*Phil. Mag.*, June, 1900).

Since $a \theta$ represents the mean kinetic energy of any gas

THEORY OF METALLIC CONDUCTION.

at the absolute temperature θ, we can calculate the value of a, and thus arrive at a numerical estimate of the amount of radiation given by the preceding expression. If we find this coincides with the observed amount it will be a strong confirmation of the theory.

By the kinetic theory of gases, if p is the pressure, N the number of molecules per unit volume of the gas—

$$p = \frac{1}{3} N\, mv^2,$$

hence $\frac{1}{2} mv^2$, the mean kinetic energy of a particle, is equal to $3p/2N$, but $\frac{1}{2} m v^2 = a\theta$, hence—

$$a\theta = \frac{3}{2}\frac{p}{N}.$$

Now at the pressure of 760 millimetres of mercury and a temperature of 0° C., $p = 10^6$, $\theta = 273$, and $N = 4 \times 10^{19}$, hence $a = 1\cdot 32 \times 10^{-16}$. Assuming that the radiation is expressed by equation (1), we can use the equation if we know the amount of radiation to find a, and Lorentz finds from the experiments made by Lummer and Pringsheim and Kurlbaum on the amount of radiation given out by hot bodies that $a = 1\cdot 2 \times 10^{-16}$. Thus the argument between theory and the results of experiment is very satisfactory and gives us considerable confidence in the truth of the theory. It ought, however, to be pointed out that we should get the same expression for the radiant energy E, whatever may be the mass or charge of the moving electrified bodies, which are supposed to generate this energy by their collisions and absorb it by their motion in the electric field, provided that the mean kinetic energy of these bodies had the same value as that we have assumed for the corpuscles.

The energy calculated in this way by Lorentz is only a part of the energy radiated in consequence of the collisions. It is that part which, when the electric forces produced by the collisions is expressed by Fourier's method as the sum of a number of harmonic components, corresponds to the part of the disturbance which can be expressed by the terms with exceedingly long wave lengths. But the disturbance, as we have seen, consists in a succession of exceedingly thin pulses, the thickness of the pulse being comparable with the distance passed over by light in the time occupied by a collision, while the part calculated by Lorentz is only the part which can be represented by harmonic terms whose wave length is long compared with the distance passed over by light, not in the short space occupied by a collision, but in the much longer interval which elapses between two collisions. It is evident that Lorentz's investigation leaves out of consideration a large part of the radiation, and that this part, arising from the accumulation of a number of thin pulses, will be analogous to the Rontgen rays—that, in fact, they will be Rontgen rays, mainly of a very absorbable type, since the corpuscles which produce them are moving much more slowly than the cathode rays in the ordinary Rontgen ray bulb. In fact, a mathematical investigation leads us to the conclusion that, of the energy radiated at a collision, there will be more of this type than the long wave type calculated by Lorentz. The character of the radiation will depend upon the time taken by a collision between the corpuscle and a molecule, if this time is so short that the distance travelled by light during the collision is very small compared with the wave length of light in the visible part of the spectrum, then the resulting radiation will be of the Rontgen ray type and not visible light. If, however, the time of collision is so prolonged that light during this time can travel over a distance comparable with the wave length of light in the visible part of the spectrum, then the resulting radiation will be visible light, and the maximum intensity of this light will be in that part of the spectrum where the wave length is comparable with the distance travelled by light during a collision, *i.e.*, when the period of vibration of the light is comparable with the time of a collision. The intensity of light having smaller wave lengths than this will rapidly fall off as the wave length diminishes. Thus in the case of these prolonged collisions

THEORY OF METALLIC CONDUCTION.

the radiation would be ordinary light, the intensity rising to a maximum at a particular part of the spectrum and then diminishing rapidly in the region of smaller wave lengths. These are characteristic properties of the radiation emitted by a black body. We know, however, the character of the radiation from such a body depends only upon the temperature and not at all upon the nature of the body, thus the colour of the light at which the intensity of the radiation is a maximum depends only on the temperature moving towards the blue end of the spectrum as the temperature is increased. On the theory that this radiation arises from the collision of corpuscles the wave length where the intensity of the radiation is a maximum depends on the duration of the collision; hence, if the radiation from hot substances arises in the way we have supposed, the duration of a collision between a corpuscle and a molecule of the substance must be independent of the nature of the substance and depend only upon the temperature, and the higher the temperature the shorter must be the duration of the collision.

By the application of the Second Law of Thermodynamics it has been shown that when the body is at the absolute temperature θ the amount of energy in the part of the spectrum comprised between wave lengths λ and $\lambda + d\lambda$ must be of the form $\lambda^{-5} \phi (\lambda \theta) d\lambda$; where ϕ is a function which cannot be determined by thermodynamical principles alone. The mathematical theory of the production of radiation by collisions shows that this energy is given by an expression of the form $\lambda^{-5} F \left(\frac{}{VT} \right) d\lambda$ where T is the duration of the collision V the velocity of light and F represents a function whose form depends upon the nature of the forces exerted during the collision. Comparing these two expressions we see that T must be conversely proportional to θ, that is, inversely proportional to the square of the velocity of the corpuscles. The velocity of corpuscles at 0° C. when in temperature equilibrium with their surroundings is about 10^7 cm./sec., the wave length at which the intensity is greatest at 0° C. is about 10^{-3} cm. In a Rontgen ray bulb giving out hard rays the velocity of the corpuscles may be about 10^{10} cm./sec., or 10^3 times the velocity of those in the metal; hence, if the law of duration of impacts is true, the radiation produced by the impact of the corpuscles in the tube should be a maximum for a wave length of $10^{-3}/10^6$ or 10^{-9} cm., as this is of the same order as the thickness of a pulse of very penetrating Rontgen radiation; this test, as far as it goes, confirms the law of the duration of collisions.

The Effect of a Magnetic Field on the Flow of an Electric Current: The "Hall Effect."

Hall found that the lines of flow of an electric current through a metallic conductor are distorted when the conductor is placed in a magnetic field. The distortion is of the character which would be produced if an additional electromotive force were to act at right angles to the original one producing the current, and also at right angles to the magnetic force. Thus if a horizontal electromotive force producing a current from right to left acts on a thin piece of metal in the plane of the paper, if the plate is placed in a magnetic field whose lines of force are at right angles to the plane of the paper and downwards, the current is distorted as if a small vertical electromotive force in the plane of the paper acted upon the metal. In some metals—for example, bismuth and silver—this force would be vertically upwards; in others, such as iron, cobalt, and tellurium, the force would be vertically downwards. In some alloys it is said that the force is in one direction for small magnetic forces and in the opposite direction for large ones. In many cases it is not proportional to the magnetic force. The theory of electric conduction we have been considering would indicate a distortion of the lines of flow of a current by a magnetic field, as the following considerations will show.

Suppose a current of electricity flows from right to left through the plate. This, on the view of the

THEORY OF METALLIC CONDUCTION. 69

current previously taken, indicates that the negative corpuscles have, on the average, a finite velocity from left to right. Let the average value of this velocity of drift of the negative corpuscles be u. If a magnetic force downwards at right angles to the plate acts on these corpuscles, they will be acted on by a vertically upward force in the plane of the paper, equal numerically to Heu, where e is the magnitude of the charge on the corpuscle, and H is the intensity of the magnetic force. The force on the corpuscle is the same as if there were an electromotive force acting vertically downwards in the plane of the paper. Thus, there would be a distortion of the lines of flow of the same sign and character as the Hall effect in bismuth. If, however, this were a complete representation of the action of the magnetic field on the current, the Hall effect would be of the same sign—the sign it has for bismuth—in all metals, and would always be proportional to the magnetic force; neither of these statements is true. Inasmuch as the Hall effect would be of the opposite sign, if the carriers of the electricity through the metal were positively charged particles instead of negatively charged ones, some physicists, in order to explain the existence of Hall effects of opposite signs, have assumed that electricity is carried through metals by two types of carriers, one positively the other negatively electrified; in some metals the negative carriers are predominant, in others the positive. There are, I think, two very serious objections to this assumption. In the first place we have no evidence of the existence of positively electrified particles able to thread their way with facility through metals, and in the second place the assumption does not explain the various phenomena connected with the Hall effect. It would indeed explain the existence of Hall effects of different signs, but on this hypothesis the amount of the Hall effect would be proportional to the magnetic force, which is by no means the case for all substances.

The complexity of the laws of the Hall effect suggests that it is due to several causes, but we can, without calling in the aid of positively charged carriers of electricity, see other sources for the variation in sign, and the failure to be directly proportional to the magnetic force. In the preceding investigation we have considered merely the effect of the magnetic force on the particle during its free path, and have neglected any influence of the magnetic force on the collisions between the corpuscles and the molecules. We can, however, easily see how a magnetic field might make suitable molecules arrange themselves so that they produce a rotatory effect on the motion of a corpuscle when the corpuscle came into collision with the molecule, and that the sign of this effect might in some cases be the same as, in others opposite to, the rotation produced by the magnetic field when the corpuscle was travelling over its free path. Thus—to take a simple instance—imagine a body whose molecules are little magnets; then if the body is placed in a magnetic field such that the lines of force are vertical and downwards, the molecules of the body will arrange themselves so that their axes tend to be vertical, the negative poles being at the top, the positive at the bottom. Then close to the magnet, in the region between its poles, the lines of force due to the magnet will be in the opposite direction to those due to the magnetic field, and the intensity of the force close in to the magnet may be very much greater than that of the external field. In this case when the corpuscle came into collision with a molecule the velocity would be rotated in the opposite direction to its rotation by the magnetic field before it came into collision with the magnet, $i.e.$, while it was travelling over its mean free path. In this case the expression for the Hall effect would consist of two terms, one arising from the free path, the other from the collisions, and these terms would be of opposite signs. If the molecules were small portions of a diamagnetic substance it is easy to see that the effect due to the collisions would be of the same sign as that due to the free path. It is perhaps worthy of note that, with the exception of tellurium, which has quite an abnormal value, the substance for which the Hall effect has the largest negative value, calling the free path effect

THEORY OF METALLIC CONDUCTION. 71

positive, is iron. It would be interesting to see if in exceedingly strong magnetic fields, much stronger than those required to saturate the iron, the Hall effect would change sign.

We must, however, I think, be careful not to import from the kinetic theory of gases ideas about the free paths of corpuscles which may not be applicable in the case of metals. The study of metals by means of micro-photography has shown that their structure is extremely complex. This is illustrated by Fig. 21, which represents the appearance under the microscope of a piece of cadmiun

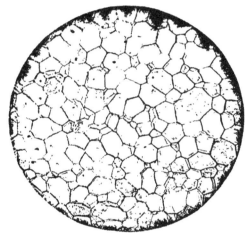

FIG. 21.

when polished and stained. A piece of metal apparently consists of an assemblage of a vast number of small crystals, and the appearance of the metal when strained past the limit of perfect elasticity shows that under strain these crystals can slip past each other. The structure of a piece of metal is thus quite distinct, from that of a gas, where the particles are distributed at equally spaced intervals. In a metal, on the other hand, it would seem that the molecules of the metal are collected in clusters, each cluster containing several molecules, and that the metal is built up of aggregates of such clusters. The collisions which determine the free path of a corpuscle may be with these clusters and not with the individual molecules, and if this were so, large variations in the free path might be brought about by variations in the number of molecules in each cluster without any variation of corresponding magnitude in the density of the metal. Thus, to take a simple case, suppose that the clusters are little spheres, and let us compare the free paths of a corpuscle (1) when there are n spheres of radius a per unit volume; and (2) when there are m spheres of radius b, the amount of matter per unit volume being the same in the two cases, so that $n\, a^3 = m\, b^3$. If λ_1 and λ_2 are respectively the free paths in the two cases, then—

$$\lambda_1 = \frac{1}{n\, \pi\, a^2} \quad \text{and} \quad \lambda_2 = \frac{1}{m\, \pi\, b^2};$$

and since $n\, a^3 = m\, b^3$ we have—

$$\lambda_1/\lambda_2 = a/b.$$

So that in this case the free path would be proportional to the radius of the cluster. Thus the bigger the cluster the longer the free path. It follows that if a rise in temperature caused the clusters to break up to some extent and become smaller, it would produce a considerable diminution in the free path of a corpuscle without any marked change in the density, whereas in a gas a rise in temperature unaccompanied by a change in density would, if the collisions between the molecules of a gas were like those between hard elastic spheres, produce no change in the free path. If the theory of conduction of electricity by corpuscles in temperature equilibrium with their surroundings is true, we must, I think, suppose that there is large variation of the free path with the temperature and with the nature of the metal. We shall see from the consideration of the Peltier effect that the number of free corpuscles per unit volume does not, in general, vary greatly from one metal to another; so that the very large variations in the electrical resistance of metals must arise much more from variations in the free paths of the corpuscles than from variations in the number of corpuscles. Hence the ratio of the free paths of the corpuscles will be of the same order as the ratio of their conductivities for electricity. Now, if the free paths of the corpuscles in the metal were determined by the same considerations as in a gas, *i.e.*, if λ were to be equal to $N/\pi a^2$, N being the number of molecules per unit volume, and a the radius of the molecules, we can show that the variations in λ would not be nearly large enough to explain the variation in the electrical conductivity. For we can determine N by dividing the density of the metal by its atomic weight, and we can get some information as to the value of a^3 from the values of the refractive indices of compounds of the different metals. Doing this, we find that the variations in $1/N \pi a^2$ are not nearly so large as the variations in the electric conductivity, and that there is little, if any, correspondence between these quantities. Moreover, if the theory we are discussing is correct there must not merely be large variations in the value of λ for the different metals, but even in the same metal at different temperatures. This follows from the consideration of the Thomson effect, *i.e.*, the convection of heat by an electric current flowing along an unequally heated conductor.

Peltier Difference of Potential between Metals.

Suppose that we place two metals A and B, which are at the same temperature, in contact, and that the pressure of the corpuscles (*i.e.*, $\frac{1}{3} N m v^2$ where N is the number of corpuscles in unit volume, m the mass, v the mean velocity of the corpuscles) in A is greater than that in B. Then corpuscles will flow from A to B; but as these corpuscles are negatively charged, the flow of corpuscles will charge B negatively and A positively. The attraction of the positive electricity in A will tend to prevent the corpuscles escaping from it, and the flow will cease when the attraction of the positive electricity in A and the repulsion of the negative in B just balances the effect of the difference in pressure. The positive electrification in A and the negative in B will be close to the surface of separation, and these two electrifications will produce a difference in electric potential between A and B, which we can calculate in the following way.

Let us suppose that there is a thin layer between the substances A B, in which the transition from A to B takes place gradually. Let N be the number of corpuscles per unit volume at a point distant x from one of the boundaries of this layer, p the pressure of the corpuscles at this point, and X the electric force. Then if e is the charge on a corpuscle, the force acting on the corpuscles per unit volume is $X Ne$. This, when there is equilibrium, must be balanced by the force arising from the variation in pressure as we pass from one side of the layer to the other. The force due to the pressure is $\frac{dp}{dx}$, hence—

$$\frac{dp}{dx} = XNe.$$

But if θ is the absolute temperature—

$$p = \frac{2}{3} a N \theta;$$

hence, if the temperature is constant across the layer, we have—

$$\frac{2}{3} a \theta \frac{1}{N} \frac{dN}{dx} = X e.$$

Integrating both sides of this equation across the layer, we get—

$$\frac{2}{3} \frac{a\theta}{e} \log \frac{N_1}{N_2} = V,$$

where V is the difference in potential between the two sides of the layer and N_1 and N_2 are the numbers of corpuscles per unit volume in A and B respectively. Thus in crossing the junction of two metals there will, unless the number of corpuscles in the two metals is the same, be a finite change in potential. Now $\frac{2}{3} a \theta/e = p/Ne$, and since it is the same for all gases we may take the case of hydrogen at 0° C. and atmospheric pressure for which $p = 10^6$, and $Ne = \cdot 41$; thus at 0° C. $\frac{2}{3} a \theta/e = 2\cdot 5 \times 10^6$, so that in volts—

The potential differences which arise in this way are not comparable with the volta differences of potential between metals in contact, for to produce a potential difference of one volt, log $N_1/N_2 = 40$, or $N_1/N_2 = 2\cdot36 \times 10^{17}$—a result quite incompatible with the comparative values of the resistances of two such metals as copper and zinc. Comparatively small variations in the number of corpuscles would, however, produce potential differences quite comparable with those measured by the Peltier effect, i.e., the heating or cooling of the junction of two metals when an electric current passes across them. Thus, to take a case where the Peltier effect is exceptionally large, that of antimony and bismuth, whose V at 0° C. is about 1/30 of a volt, we see from equation (1) that for these metals log $(N_1/N_2) = 1\cdot33$, or $N_1/N_2 = 3\cdot8$. Thus, if the number of corpuscles in the unit volume of antimony were about four times that in bismuth we should, on this theory, get Peltier effects of about the right amount. Since the Peltier effect for antimony and bismuth is very much larger than that for most pairs of metals, we see that the theory indicates that in general the number of free corpuscles per unit volume does not vary much from one metal to another. From the Peltier effects of each metal with a standard metal we can get the ratio of the number of corpuscles in these metals to the number in the standard metal. Having done this, since at the same temperature the conductivity of the metals is proportional to the product of the number of corpuscles per unit volume and the free path of a corpuscle in the metal, we can get the ratio of the free paths in the different metals, and we can then see whether the free paths obtained in this way can be reconciled with the other properties of the metals. The result of such a comparison leads, I think, to the conclusion that the mechanism by which we have supposed the electric current to be conveyed through a conductor is at most only a part and not the whole of the process of metallic conduction. One reason for this conclusion is the large changes which take place in the electrical resistance of some metals at fusion, changes which do not seem to be accompanied by any corresponding change in their thermo-electric quality. Thus the conductivities of tin, zinc and lead at their melting points are, when the metals are in the solid state, about twice what they are in the liquid. These metals all contract on solidification, so that the average distance between the molecules is greater in the liquid than in the solid state. The electrical conductivity varies as the product of N the number of corpuscles per unit volume, and λ the free path of a corpuscle. Since the distance between the molecules is greater in the liquid than in the solid state, we should expect the free path of the corpuscles to be greater, but if $N_1 \lambda_1$ and $N_2 \lambda_2$ are respectively the values of $N \lambda$ in the solid and liquid states, $N_1 \lambda_1 = 2 N_2 \lambda_2$, and since λ_2 is greater than λ_1, N_1 must be greater than $2 N_2$. A reference to equation (1) will show that this involves a Peltier effect between the solid and the liquid metal of about half the magnitude of that between bismuth and antimony, and thus, as these effects go, exceedingly large. Now Fitzgerald, Minarelli and Obermeyer, as quoted by G. Wiedemann, "Elektricität," ii., p. 289, could detect no sudden change in thermo-electric circuits with these metals when they passed from the solid to the liquid state, whereas if the number of free corpuscles had diminished to one half, the effect would have been very conspicuous. There is thus a discrepancy between the results of the determination of the relative number of corpuscles in the two states by data derived (1) from thermo-electric phenomena; (2) from their electric resistance. This discrepancy is so large that it is impossible to suppose it is due to any errors in the data derived from experiment.

The Thomson Effect.

Lord Kelvin showed that in some metals an electric current carries heat from the hot to the cold parts of the metal, while in other metals the transference of heat is in the opposite direction. Let us calculate what this transference of heat would be on the theory we are discussing.

THEORY OF METALLIC CONDUCTION. 77

Let AB be a bar of metal, and let the temperature increase from A to B. If the pressure of the corpuscles depends upon the temperature there must be electromotive forces along the bar to keep the corpuscles from drifting under these pressure differences. If p is the pressure of the corpuscle at a point distant x from the end A, then the force acting on the corpuscles included between two planes at distances x, $x + \Delta x$, from A, is, per unit area of these planes, equal to $\Delta x \dfrac{dp}{dx}$ and acts from right to left. To balance this we must have an electromotive force X tending to move the corpuscles from left to right, determined by the equation—

$$X e n \Delta x = \frac{dp}{dx} \Delta x$$

or—

$$X e = \frac{1}{n} \frac{dp}{dx},$$

where n is the number of corpuscles per unit volume at a distance x from A. If θ is the absolute temperature of the bar at A we have (see page 65)—

$$p = \frac{2}{3} n a \theta.$$

hence—

$$X e = \frac{2}{3} \frac{1}{n} \frac{d}{dx} (a n \theta).$$

Hence a corpuscle in travelling from $x + \delta x$ to x will abstract from the metal an amount of heat whose mechanical equivalent is $X e \, \delta x$, or—

$$\frac{2}{3} \frac{1}{n} \frac{d}{dx} (a n \theta) \, dx.$$

The corpuscle when at $x + dx$ has an amount of kinetic energy equal to $a \left(\theta + \dfrac{d\theta}{dx} dx \right)$, while at x its kinetic energy is reduced to $a \theta$, hence the corpuscle will communicate to the metal between x and $x + dx$ an amount of heat equal to $a \dfrac{d\theta}{dx} dx$; thus the total amount of heat communicated by the corpuscle to the metal is—

$$\left\{ a \dfrac{d\theta}{dx} - \dfrac{2}{3} \dfrac{1}{n} \dfrac{d}{dx}(a\,n\,\theta) \right\} dx,$$

or—

$$\left(a - \dfrac{2}{3} \dfrac{1}{n} \dfrac{d}{d\theta}(a\,n\,\theta) \right) d\theta.$$

If the current i is flowing in the direction in which x increases, the number of corpuscles which cross, unit area in unit time, in the opposite direction to the current is i/e, and the mechanical equivalent of the heat they communicate to the metal between the places where the temperatures of the metal are respectively θ and $\theta + d\theta$ is equal to—

$$\dfrac{i}{e}\left(a - \dfrac{2}{3} \dfrac{1}{n} \dfrac{d}{d\theta}(a\,n\,\theta) \right) d\theta.$$

But if σ is the "specific heat of electricity in the metal," this amount of heat is by definition equal to—

$$- i\,\sigma\,d\theta$$

the minus sign being inserted because the current is flowing from the cold to the hot part of the circuit; hence—

$$\sigma = - \dfrac{1}{e} \left(a - \dfrac{2}{3} \dfrac{1}{n} \dfrac{d}{d\theta}(a\,n\,\theta) \right)$$

$$= - \dfrac{a}{3e} \left\{ 1 - 2\theta \dfrac{d}{d\theta} \log n \right\} \tag{1}$$

$$= \dfrac{2}{3} \dfrac{a}{e} \theta \dfrac{d}{d\theta} \log n\,\theta^{-\frac{1}{2}} \tag{2}$$

The term $\dfrac{-}{3}$ in the expression for σ is the same for all metals, and since the electro-motive force round a thermo-electric circuit consisting of two metals only involves the *difference* of the specific heats of electricity in the metals, this term will not affect the electromotive force round the

THEORY OF METALLIC CONDUCTION. 79

circuit. It will, however, affect the amount of heat developed in the conductor, and we shall find that unless this term is very nearly balanced by the term $\dfrac{a}{3} - \dfrac{i}{a\nu}\log n$, the amount of heat developed by the flow of a current through an unequally heated conductor would be far greater than the amount actually observed.

For $a/3e$ is about 0.45×10^4, so that the amount of heat expressed by the first term in equation (1) developed by a unit current in flowing between two places where the temperature differed by 1° C. would equal $.45 \times 10^4/4.2 \times 10^7$, or 1.07×10^{-4} calories per second.

The metal in which this heat effect is largest is, as far as our present knowledge extends, bismuth, and for this metal the observed effect is only about $.3 \times 10^{-4}$ calories, or about 1/3 of the amount expressed by the term $a/3\,e$, and the effect in bismuth is very much greater than in any other metal; hence since σ is small compared with $a/3\,e$, we have by equation (1)

$$\log n - \tfrac{1}{2} \log \theta + \text{a constant}$$

approximately, so that approximately n will vary as $\theta^{\frac{1}{2}}$, i.e., the number of free corpuscles will vary approximately as the square root of the absolute temperature. If the specific heat of electricity is positive the number of free corpuscles will vary a little more rapidly than this with the temperature. If the specific heat is negative it will vary a little less rapidly. This variation of the number of free corpuscles with the temperature involves a still more rapid variation of the mean free path. For (see p. 54) we have seen that the electrical conductivity is proportional to $n \lambda v / \theta$. Now v is proportional to $\theta^{\frac{1}{2}}$ and n, as we have just seen, varies approximately according to the same law, hence the electrical conductivity is approximately proportional to λ the free path of the corpuscles in the metal. But for many pure metals the electrical conductivity varies approximately as the reciprocal of the absolute temperature; hence for these metals the mean free path must also vary with the temperature in the same way, *i.e.*, be inversely proportional to the absolute temperature. This rapid variation of the free path with the temperature would not be possible if the structure of the metal were analogous to that of a gas compressed so that the distances between the molecules were all diminished in the same proportion. We have seen that if the metal consisted of aggregations of molecules which broke up to some extent as the temperature rose, we might get a rapid variation of the mean free path, with the temperature. Since the free path, according to this theory, varies approximately as the reciprocal of the absolute temperature, the free paths at the low temperatures which can be obtained by the use of liquid air or liquid hydrogen ought to be much greater than at ordinary laboratory temperatures. Thus the effects which depend on the free path, such as the effect of magnetic force on electrical resistance, or the absorption of light by the metal (which should vary greatly according as the time of vibration of the light is greater or less than the time occupied by a corpuscle to describe its free path), would be greatly affected by the lowering of the temperature: experiments on these points would be valuable tests of the theory. If λ varies as $1/\theta$, λ/v the time occupied by a corpuscle in describing its free path will vary as $1/\theta^2$. The velocity acquired by a corpuscle under a constant electric force will also vary as $1/\theta^{\frac{3}{2}}$, and will thus diminish rapidly as the temperature increases.

The Number of Free Corpuscles in Unit Volume of the Metal.

We can determine from the amount of heat absorbed or developed when a current of electricity passes across the junction of two metals, the ratio of the number of corpuscles in unit volume of the two metals, and from the Thomson effect we can determine the change in this number for any one metal with the temperature. Hence,

THEORY OF METALLIC CONDUCTION. 81

if we can determine the number of corpuscles per unit volume in any one metal at any one temperature, we can deduce the number in any other metal at any temperature.

We shall now pass on to the consideration of methods to determine the absolute number of corpuscles per unit volume; since the electrical conductivity gives us the value of $n\lambda$, a method of determining λ will also lead to the determination of n. We shall begin with those methods which lead to the direct determination of n.

One of the simplest of these in principle is founded on the consideration of what takes place when a charge of electricity is communicated to a piece of metal. Let us, to fix our ideas, suppose that the charge is a negative one and that it is carried by free corpuscles. These corpuscles must occupy a layer of finite thickness at the surface of the metal, for if the layer were reduced to infinitesimal thickness the pressure exerted by these corpuscles would be vastly greater than the pressure exerted by the corpuscles in the interior of the metal, and the consequence would be that corpuscles would diffuse from the layer into the interior of the metal. The corpuscles will diffuse until the electric force exerted by their charges is just able to balance the forces arising from the difference of pressure between the surface and the interior. We can calculate the thickness of the layer occupied by the negative charge in the following way: Let A be the face of a flat piece of metal having a negative charge; let n be the number of corpuscles per unit volume before the charge was communicated to the metal, $n + \xi$ the number at a point at a distance x from the surface of the plate after the charge was communicated, p the pressure of the corpuscles at this distance, and X the electric force tending to stop the corpuscles from moving from left to right. Then when the corpuscles have got into a steady state—

$$\frac{dp}{dx} = Xe(n + \xi),$$

but $p = \frac{2}{3} a (n + \xi) \theta$, where $a \theta$ is the mean kinetic

T.M. G

energy of a corpuscle at the absolute temperature θ, and since n does not depend upon x, we have, assuming that ξ is small compared with n—

$$\frac{2}{3} \theta \cdot \frac{d\xi}{dx} = X e n;$$

but—

$$\frac{dX}{dx} = 4\pi \xi e,$$

if e is measured in electrostatic units, hence—

$$\frac{2}{3} u v \frac{d^2\xi}{dx^2} = 4\pi e^2 n \xi,$$

or—

$$\xi = A \epsilon^{-px}$$

where $p^2 = \frac{4\pi e^2 n}{\frac{2}{3} a \theta}$ and A is a constant. To find A we have $e \int_0^\infty \xi\, dx = Q$, if Q is the charge per unit area; hence substituting for ξ, $\frac{eA}{p} = Q$, or

$$\xi = \frac{pQ}{e} \epsilon^{-px}.$$

Thus the value of ξ is appreciable until x is large compared with $1/p$; we may thus take $1/p$ or $(a\theta/6\pi e^2 n)^{\frac{1}{2}}$ as the measure of the thickness of the layer occupied by the electricity; substituting for $a\theta$ and e the values $3{\cdot}6 \times 10^{-14}$ and 3×10^{-10}, we find that, at $0°$ C.,

$$\frac{1}{p} = \{10^6/15\pi n\}^{\frac{1}{2}}$$

Now since—

$$\frac{dX}{dx} = 4\pi e \xi$$

we have—

$$X = 4\pi Q \epsilon^{-px}$$

and—

$$\int_0^\infty X\, dx = \frac{4\pi Q}{p}.$$

This is the difference in potential between the surface and a point in the interior, hence we see that if we communicate a charge of electricity to a hollow conductor whose surface

THEORY OF METALLIC CONDUCTION.

is kept at zero potential, the interior of that conductor will not, as is usually assumed in electrostatics, remain at zero potential, but will change by $4\pi Q/p$ where Q is the charge per unit area of the conductor. Hence, if we measure the change produced by a known charge we shall determine p and hence n by the equation $15\pi n = 10^6 p^2$. If the number of corpuscles is comparable with the number of molecules of the metal, which we may take as between 10^{22} and 10^{23}, p will be comparable with 10^8, and so the thickness of the layer through which the electricity is distributed will be of the order of 10^{-8} cm. In this case the change in the potential of the interior produced by any feasible charge will be small, but not perhaps too small to be measurable. If the conductor were exposed to air at atmospheric pressure the greatest value of $4\pi Q$ possible without sparking would be 100 in electrostatic measure. By embedding the conductor in a solid dielectric, such as paraffin, we could probably increase $4\pi Q$ to 1000 without discharge. If $4\pi Q$ is 10^3 and $p = 10^8$, the change in potential would be 10^{-5} in electrostatic measure, or 3×10^{-3} of a volt, and this ought to be capable of measurement.

Experiments have been made by Bose and others to see if the electrical resistance would be altered by giving a charge of electricity to a very thin conductor; so far these have led to negative results. We might at first sight expect that if we increased the supply of negative corpuscles by communicating a charge of negative electricity to the strip of metal we should increase the conductivity; but this need not necessarily be the case, for suppose the surface instead of being flat were corrugated, then the charge would be all at the tops of the corrugations; but this would be quite out of the way of a current flowing through the film, which would take the short circuit through the base of the corrugations. As the electricity only penetrates a distance comparable with the size of a molecule, it is impossible to avoid an effect of this kind, however carefully the surface is polished.

We can, however, find both lower and upper limits to the number of free corpuscles, and as these limits lead to

84 THE CORPUSCULAR THEORY OF MATTER.

contradiction we shall, after investigating them, proceed to the consideration of the question whether the other view of the function and disposition of the corpuscles alluded to on page 49 is less open to objection.

We can obtain a lower limit to the number of free corpuscles per unit volume of a metal by the consideration of the results of the experiments of Rubens and Hagen on the reflection of long waves from the surface of metals. It follows from these experiments that the electrical conductivity of metals when waves whose length equals 25 μ, μ being 10^{-4} cm., pass through them is the same as the conductivity under steady electrical forces, and that even when the waves are as short as 4 μ the electrical conductivity is within about 20 per cent. of that for steady forces. We can easily show that if k is the conductivity under steady forces; then when the forces vary as $\sin n\,t$ the conductivity will be proportional to $k\,\dfrac{\sin^2 n\,T}{2\,n^2 T^2}$, where $2\,T$ is the interval between two collisions. Thus, unless this interval be small compared with the period of the electric force the conductivity will be very materially reduced. Thus if T were as great as one quarter of the period of the force, so that $n\,T = \dfrac{\pi}{2}$, the conductivity would be reduced to $1/(\pi/2)^2$, or ·4 of its steady value. As the diminution of the conductivity for light waves whose length is 4 μ is less than this, we conclude that the interval between two collisions is less than one-quarter the period of this light, or less than $3\cdot 3 \times 10^{-15}$ sec. Hence u the velocity under unit electric force, since it is equal to $\dfrac{1}{2}\dfrac{e}{m}\,T$, will be less than $\tfrac{1}{2}\,3\cdot 3 \times 10^{-15}\dfrac{e}{m}$, and since k the conductivity is $n\,e\,u$, n will be greater than k/eu, i.e., than $\dfrac{k\,10^{15}\,m}{1\cdot 6\,e^2}$.

For silver k is about 5×10^{-4}, and since $e/m = 1\cdot 7 \times 10^7$ and $e = 10^{-20}$, we see that n for this metal must be greater than $1\cdot 8 \times 10^{24}$.

THEORY OF METALLIC CONDUCTION. 85

It is this result which leads to the difficulty to which we have alluded, for if there were this number of corpuscles per unit volume, then, since the energy possessed by each corpuscle at the temperature θ is $a\,\theta$, the energy required to raise the temperature of the corpuscles in unit volume of the metal by 1° C. is $n\,a$, and since $a = 1.5 \times 10^{-16}$ (see page 65), the energy which would have to be communicated to unit volume of the silver to raise the temperature of the corpuscles alone would be greater than $1.3 \times 1.8 \times 10^8$ ergs., or about 6 gram calories. But to raise the temperature of a cubic centimetre of silver one degree, only requires about 0·6 calories, and this includes the energy required to raise the temperature of the atoms of the metal as well as that of the corpuscles. We thus get to a contradiction. The value of the specific heats of the metals shows that the corpuscles cannot exceed a certain number, but this number is far too small to produce the observed conductivities if the intervals between the collisions are as small as is required by the behaviour of the metals in Rubens' experiments.

CHAPTER V.

THE SECOND THEORY OF ELECTRICAL CONDUCTION.

WE shall now proceed to develop the second theory of electrical conductivity and see whether it is as successful in explaining the relation between the thermal and electrical conductivities as the other one, and whether or not it is open to the same objections.

On this theory the corpuscles are supposed to be pulled out of the atoms of the metal by the action of the surrounding atoms. In order to get a sufficiently definite idea of this process to enable us to calculate the amount of electrical

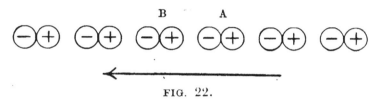

FIG. 22.

conductivity which it would produce, we shall suppose that in the metal there is a large number of doublets, formed by the union of a positively electrified atom with a negatively electrified one, and that the interchange of corpuscles takes place by a corpuscle leaving the negative component of one of these doublets and going to the positive constituent of the other. Under the action of the electric force these doublets tend to arrange themselves along that line in the way indicated in Fig. 22, much in the same way as the Grotthus chains in the old theory of electrolysis. The corpuscles moving in the direction of the arrows will give rise to a drift of negative electricity against the direction of the electric force or a current of positive electricity in the same direction as the force.

We now proceed to calculate the magnitude of the current

produced in this way. Consider a doublet formed by a charge of electricity $+e$, connected with another charge $-e$, and placed in an electric field where the intensity of the electric force is X. The potential energy of the doublet, when its axis (the line joining the negative to the positive charge) makes an angle θ with the direction of the electric force, is $-Xed\cos\theta$, where d is the distance between the charges in the doublet. If the doublets distribute themselves as they would in a gas in which the distribution of potential energy follows Maxwell's law, the number possessing potential energy V will be proportional to ϵ^{-hV}, where $1/h = \frac{2}{3}\alpha\theta$, $\alpha\theta$ being as before the mean kinetic of a molecule at the absolute temperature θ. Then the number of doublets whose axes make an angle between θ and $\theta + d\theta$ with the direction of X, is proportional to $\epsilon^{hXed\cos\theta}\sin\theta\,d\theta$, and the average value of $\cos\theta$ for these doublets is equal to

$$\frac{\int_0^\pi \epsilon^{hXed\cos\theta} \cos\theta \sin\theta\,d\theta}{\int_0^\pi \epsilon^{hXed\cos\theta} \sin\theta\,d\theta.}$$

Now $Xedh$ will, unless the electric force greatly exceeds the value it has in any ordinary case of metallic conduction, be exceedingly small, for the potential difference through which the charge e must fall in order to acquire the energy possessed by a molecule at the temperature 0° C., is about 1/25 of a volt, and h is proportional to the reciprocal of this energy, thus unless the electric field is so strong that there is in the space between the two components of the doublet a fall of potential comparable with this, $hXed$ will be small. But when this is so—

$$\int_0^\pi \epsilon^{hXed\cos\theta} \cos\theta \sin\theta\,d\theta = \frac{2}{3} hXed$$

and—
$$\int_0^\pi \epsilon^{hXed\cos\theta} \sin\theta\,d\theta = 2,$$

hence the mean value of $\cos\theta$ is $\frac{1}{3} hXed$, or $\frac{2}{\alpha}\frac{Xed}{\theta}$.

If each doublet discharges a corpuscle p times a second, then in consequence of the polarisation we have just investigated, the resultant flow of corpuscles will be the same as if each doublet discharged a corpuscle parallel but in the opposite direction to the electric force $p \times \dfrac{2}{9} \dfrac{X d}{a \theta}$ times per second.

Hence, if n is the number of doublets per unit volume, b the distance between the centres of the doublets, the current through unit area will be equal to—

$$\frac{2}{9} \frac{e X d p n b}{a \theta}.$$

If we assume that the orientation of the axes of the doublets in a metal follows the same law as in a gas, this will be the expression for the current through the metal, hence c the electrical conductivity will be given by the expression—

$$c = \frac{2}{9} \frac{e^2 d p n b}{a \theta}.$$

Thermal Conductivity.

If we suppose that the kinetic energy of the corpuscle in a doublet is proportional to the kinetic energy, *i.e.*, to the temperature of the doublet, the interchange of corpuscles will carry heat from the hot parts of the metal to the cold, and will thus give rise to the conduction of heat. Let us suppose that the kinetic energy of a corpuscle when in a doublet at temperature θ is $a\,\theta$. If the corpuscle goes from a doublet where the temperature is $\theta + \delta\theta$ to one where the temperature is θ, it will, when the latter doublet has lost a corpuscle to make way for the one coming, have caused a transference of heat equal to $a\,\delta\theta$. Consider now the transference of heat across a plane at right angles to the temperature gradient. The number of corpuscles crossing this plane in unit time is equal to $\tfrac{1}{3} n b \cdot p$. If the difference of temperature between the adjacent doublets is $\delta\theta$, this will transfer

$$\frac{1}{3} n b p a \delta \theta$$

THEORY OF ELECTRICAL CONDUCTION. 89

units of heat across the plane in unit time, but as b is the distance between the doublets $\delta\theta = \dfrac{d\theta}{dx} b$, where x is measured in the direction of the flow of heat. Hence κ the thermal conductivity is given by the equation—

$$\kappa = \frac{1}{3} n b^2 p a.$$

Thus on this theory κ/c, the ratio of the thermal to the electrical conductivity is equal to—

$$\frac{3}{2} \frac{b}{d} \frac{a^2 \theta}{e^2}.$$

On the theory discussed before this ratio was equal to—

$$\frac{4}{3} \frac{a^2 \theta}{e^2}.$$

In a substance in which the doublets are so numerous as to be almost in contact, d and b will be very nearly equal to each other, and in this case the ratio of the conductivities on the new theory would be to that on the old in the proportion of 9 to 8. When the doublets are more sparsely disseminated b will be greater than d and the ratio of the conductivities given by the new theory will be greater than that given by the old. The agreement between theory and the results of experiment is at least as good in the new theory as in the old, for the new theory gives for good conductors results of the right order of magnitude, while the presence of the factor b/d indicates that the ratio is not an absolute constant for all substances but varies within small limits for good conductors and wider ones for bad ones. All this is in agreement with experience.

THEORY OF CONNECTION BETWEEN RADIANT ENERGY AND THE TEMPERATURE.

We have seen (p. 61) that Lorentz has shown that the long wave radiation can be regarded as a part of the electromagnetic pulses emitted when the moving corpuscles come into collision with the atoms of the substance through which they are moving, and he has given an expression for the amount of the energy calculated on this principle, which agrees well with that found by experiment. But in the new theory, as in the old, we have the sudden starting and stopping of charged corpuscles and therefore the incessant production of electromagnetic pulses; these when resolved by the aid of Fourier's theorem will be represented by a series of waves, having all possible wave lengths from zero to infinity. We must see if the energy in the long wave length radiation at a given temperature would on the new theory be approximately equal to that on the old.

It will be necessary to examine a little more closely than we have hitherto done the theory of the radiation from metals due to the stopping and starting of electrified systems inside the metal. We have already (see p. 64) quoted an expression due to Lorentz for the amount of the very long wave length radiation due to the stopping of corpuscles. We can, however, by the following method, obtain an expression for the energy corresponding to any wave lengths emitted by unit volume of the metal. In the case of very long waves this expression coincides with that given by Lorentz.

We have seen that when the motion of an electrified particle is accelerated it gives off pulses of electric and magnetic force. If f is the acceleration of a charged body O, at the time t, the magnetic force at a point P at a time $t + \frac{OP}{c}$, is $\frac{e f}{c} \frac{\sin \theta}{OP}$ where θ is the angle OP makes the direction of the acceleration, and c the velocity of light. The energy per unit volume at P due to this magnetic field is equal to $\frac{H^2}{8\pi}$ where $H = \frac{e f \sin \theta}{c \; OP}$, and the amount of this energy, which flows out radially through unit area at P, is $c H^2 / 8 \pi$. Integrating over the surface of the sphere with centre O and radius OP we find that the flow of energy

THEORY OF ELECTRICAL CONDUCTION. 91

due to the magnetic field is, in unit time $\dfrac{1}{3}\dfrac{e^2 f^2}{v}$. There is an equal flow of energy due to the electric field, hence the rate at which the charged body is radiating energy is $\dfrac{2}{3}\dfrac{e^2 f^2}{v}$, a result first given by Larmor.

The total amount of energy radiated is

$$\frac{2}{3}\frac{e^2}{c}\int_{-\infty}^{+\infty} f^2\,dt.$$

When we know f as a function of t we can find the total amount of energy radiated. If we wish to find how much of this energy corresponds to light between assigned limits of wave length we must express, f by Fourier's theorem, in terms of an harmonic function of the time.

Let us take the following case as representing the stopping and starting of a charged particle in a solid. The particle starts from rest, for a time t_1 has a uniform acceleration β, at the end of this time it has got up speed and now moves for a time t_2 with uniform velocity, at the end of this time it comes into collision, and we suppose that now an acceleration $-\beta$ acts for a time t_1 and reduces it to rest again,

Thus f when expressed as a function of the time, if the time $t = o$ is taken as the time when it is at the middle of its free path, has the following values—

$f = o$ from $t = -\infty$ to $t = -\left(t_1 + \dfrac{t_2}{2}\right)$

$f = \beta$ from $t = -\left(t_1 + \dfrac{t_2}{2}\right)$ to $t = -\dfrac{t_1}{2}$

$f = o$ from $t = -\dfrac{t_1}{2}$ to $t = \dfrac{t_2}{2}$

$f = -\beta$ from $t = \dfrac{t_2}{2}$ to $t = t_1 + \dfrac{t_2}{2}$

$f = o$ from $t = t_1 + \dfrac{t_1}{2}$ to $t = \infty$

92 THE CORPUSCULAR THEORY OF MATTER.

Now by Fourier's theorem we have, if $\phi(t)$ is a function of t,

$$\phi(t) = \frac{1}{\pi} \int_0^\infty \int_{-\infty}^{+\infty} \phi(u) \cos q (u - t) \, dq \, du$$

applying this to our case, and performing the integrations, we find

$$f = \frac{4\beta}{\pi} \int_0^\infty \frac{\sin q \frac{t_1}{2} \sin q \frac{t_2}{2} \sin q \frac{(t - t_1)}{2}}{q} \sin q t \cdot dq.$$

Now Lord Rayleigh has shown (*Philosophical Magazine*, June, 1889, p. 466) that if

$$\phi(t) = \frac{1}{\pi} \int_0^\infty f_1(q) \sin q t \cdot dq$$

$$\int_{-\infty}^{+\infty} (\phi(t))^2 \, dt = \frac{1}{\pi} \int (f_1(q))^2 \, dq,$$

hence—

$$\int_{-\infty}^{+\infty} f^2 \, dt = \frac{16 \beta^2}{\pi} \int_0^\infty \frac{\sin^2 q \frac{t_1}{2} \cdot \sin^2 q \frac{(t_1 + t_2)}{2}}{q^2} \, dq$$

The energy radiated from the charged body is equal to

$$\frac{2}{3} \frac{e}{c} \int_{-\infty}^{+\infty} f^2 \, dt$$

$$= \frac{32}{3} \frac{e^2}{\pi c} \beta^2 \int_0^\infty \frac{\sin^2 q \frac{t_1}{2} \sin^2 q \frac{(t_1 + t_2)}{2}}{q^2} \, dq,$$

hence if there are s collisions per unit volume per second the energy radiated from unit volume per second is

$$s \times \frac{32}{3} \frac{e^2}{\pi c} \beta^2 \int_0^\infty \frac{\sin^2 q \frac{t_1}{2} \sin^2 q \frac{(t_1 + t_2)}{2}}{q^2} \, dq,$$

and the energy corresponding to waves which have a frequency between q and $q + dq$ is equal to

THEORY OF ELECTRICAL CONDUCTION. 93

In the case considered by Lorentz the waves are very long, i.e., q is small compared with $1/t_1$, or $1/(t_1 + t_2)$ and $s = \dfrac{n}{\lambda}$; in this case the preceding expression reduces to

$$\frac{n\,v}{\lambda}\frac{2}{3}\frac{e^2}{\pi\,c}\beta^2\,t_1^2\,(t_1 + t_2)^2\,q^2\,d\,q. \tag{B}$$

Now $\beta = v/t_1$, and if t_1, i.e., the time occupied by the collisions is small compared with t_2 the time spent in describing the free path, $\lambda = v\,t_2$, so that the preceding expressions become

$$n\,v\,\frac{2}{3}\frac{e^2}{\pi\,c}\,\lambda\,q^2\,d\,q.$$

Now k, the electric conductivity, $= \dfrac{1}{4}\dfrac{n e^2}{a\,\bar{v}}$, so that the energy radiated from unit volume in unit time is

$$\frac{8\,a}{3}\frac{\theta}{\pi\,c}\,k\,q^2\,dq.$$

We can get an expression for the stream of radiant energy by using the principle that when things have got into a steady state, the amount of energy absorbed by unit volume in unit time is equal to the energy radiated from that volume in the same time. If E is the electric force in the stream of radiant energy i the intensity of the current, the energy absorbed in unit volume per unit time is $E\,i$, or, $k\,E^2$ since $i = k\,E$. Now W, the energy per unit volume, is equal to $\dfrac{K\,E^2}{4\,\pi}$ where K is the specific inductive capacity in electromagnetic units; hence the rate at which energy is absorbed is $\dfrac{4\,\pi\,k}{K}\,W$, and this, when things are in a steady state, must be equal to the energy radiated, hence we have

94 THE CORPUSCULAR THEORY OF MATTER.

the energy in the stream of radiant energy due to waves having a frequency between q and $q + dq$ is equal to

$$\frac{2}{3}\frac{a\theta K}{\pi^4 c}q^2\,dq.$$

If μ is the refractive index of the substance

$$K = \mu^2/c^2,$$

hence the density of the stream of radiant energy is

$$\frac{2}{3}\frac{a\theta\mu^2}{\pi^2 c^3}q^2\,dq,$$

a result which Lorentz has shown agrees well with the actual determinations of the radiation. We must remember that this result only holds when the frequency of the waves is very small, not merely because it is only in this case that the expression A reduces to B, but also because when the frequency is large the conductivity k will not have the value we have assigned to it.

To return to the expression A for the amount of energy radiated. We see that the maximum amount of the energy for a given difference of frequency will be when the frequency is such that qt_1 is small and $q(t_1 + t_2)$ finite, *i.e.*, when the time of vibration of the light is comparable with the time occupied in running over the free path: the energy in the light with this frequency is greater than in the light whose frequency is very small; we can, however, easily show that, as we should expect, the greatest amount of energy is in the waves whose time of vibration is comparable with t_1, the time occupied by a collision.

We can see this as follows;—since the rate of radiation of energy is $\frac{2}{3}\frac{e^2 f^2}{c^3}$, then U the amount radiated by one corpuscle in the case we have considered is—

When the frequency is very small, the energy radiated having a frequency between q and $q + dq$ is

$$\frac{2}{5}\frac{e^2}{c\pi}\beta^2 t_1^2 (t_1 + t_2)^2 q^2 dq,$$

or U_1 the total energy of waves having frequencies between o and q is given by the equation

$$\frac{2}{9}\frac{e^2}{c\pi}\beta^2 t_1 \cdot q\, t_1 \cdot q^2 (t_1 + t_2)^2,$$

as both qt_1 and $q(t_1 + t_2)$ are very small, U_1 is only a small fraction of U, the total amount of energy radiated.

Next take the case when $q\, t_1$ is very small and $q(t_1 + t_2)$ finite; in this case the energy between q and $q + dq$ is

$$\frac{8}{3}\frac{e^2}{c\pi}\beta^2 t_1^2 \sin^2 \frac{q(t_1 + t_2)}{2} dq,$$

and U_2 the value of this over a range q is given by the equation—

$$U_2 = \frac{4}{3}\frac{}{c\pi}\beta^2 t_1 \cdot q\, t_1.$$

Since $q\, t_1$ is small, we see that U_2 is small compared with U; it is, however, large compared with U_1, the long wave thermal radiation. Since U_1 and U_2 are each small compared with U, we see that by far the larger part of the energy will be in the light whose time of vibration is of the same order as the time occupied by a collision. If this time depends on the temperature diminishing as the temperature increases, and if at a certain temperature it was of the same order as the time of vibration of visible light, the radiation at that temperature would be mainly visible light and the higher the temperature the bluer the light.

The assumptions we made as to the nature of the acceleration of the charged body, that it was equal to β for a short interval t_1, then equal to 0 for a time equal to t_2, and then equal to $-\beta$ for a time t_1, is perhaps more appropriate to the first theory of metallic conduction than to the second, where we suppose the charged body pulled out of an atom by the attraction of B, and being suddenly stopped when it strikes against B. For this case it is more appropriate to suppose that f is equal to β_2 for a time t_2, and then equal to $-\beta_1$ for a very short time t_1, where $\beta_2 t_2 = \beta_1 t_1$.

By applying the same method as before we can easily show that the energy of light with frequency between q and $q + dq$ radiated in unit time from unit volume is—

$$s \frac{2}{3} \frac{e^2}{\pi c} \left(\beta_2^2 \sin^2 \frac{q_2 t_2}{2} + \beta_1^2 \sin^2 \frac{q t_1}{2} \right.$$
$$\left. - 2 \beta_1 \beta_2 \sin \frac{q t_2}{2} \sin \frac{q t_1}{2} \cos q (t_1 + t_2) \right) \frac{dq}{q^2},$$

or

$$s \frac{8}{3} \frac{e^2}{\pi c} \beta_1^2 t_1^2 \left(\frac{\sin^2 \frac{q t_1}{2}}{q^2 t_1^2} + \frac{\sin^2 \frac{q t_2}{2}}{q^2 t_2^2} \right.$$
$$\left. - \frac{2 \sin \frac{q t_1}{2}}{q t_1} \cdot \frac{\sin \frac{q t_2}{2}}{q t_2} \cos q (t_1 + t_2) \right) dq.$$

For very small values of q this reduces to

$$s \frac{1}{6} \frac{e^2}{\pi c} \beta_1^2 t_1^2 (t_1 + t_2)^2 q^2 d q ;$$

or, since t_1 is small compared with t_2,

$$\frac{1}{6} s \frac{e^2}{\pi c} \beta_2^2 t_2^4 q^2 d q.$$

Now $\frac{1}{2} \beta_2 t_2^2$ is the space passed over by the charged body while its motion is being accelerated; it is, therefore, equal to b, the distance separating the systems between which the charged body travels, hence the energy radiated from unit volume per second is—

$$\frac{2}{3} \frac{s e^2}{\pi c} b^2 q^2 d q.$$

If W is the stream of radiant energy between these limits of frequency flowing through the body we have as before

$$\frac{4 \pi k c^2}{\ldots} W = \frac{2 s e^2}{\ldots} \ldots$$

where k is the electrical conductivity, and on this theory

$$k = \frac{2}{9} \frac{c^2 \rho n b d}{a \theta}.$$

Now $s = n \mu^2$, hence

$$W = \frac{3}{4} \frac{a \theta \rho b}{\pi^2 c^3 d} q^2 d q,$$

while on the other theory it was (see page 94) given by the equation—

$$W = \frac{2}{3} \frac{a \theta \mu^2}{\pi^2 c^3} q^2 d q.$$

If, as we should expect in a good conductor, b is very nearly equal to d, the radiation on the new theory is to that on the old as 9 to 8. Thus the expressions are so nearly equal that in the present state of our knowledge we cannot say that in this respect the one theory agrees better with the facts than the other.

On this theory by far the greater part of the radiation which starts from the metal is of exceedingly short wave length, the time of vibration being comparable with t_1.

Peltier and Thomson Effects.

These effects on the theory first discussed, that corpuscles are distributed throughout the metal and are in temperature equilibrium with it, may be regarded as arising in the following way. If there are n corpuscles per unit volume and v is their average velocity, then through unit area in the metal, $\frac{1}{6} n v$ corpuscles will in one second pass through in one direction. Hence if we have two metals A and B in contact, and if $n v$ in A is not the same as in B, the number of corpuscles that flow from A to B will not be the same as the number that flow from B to A. To fix our ideas, let us suppose that the flow through A is greater than that through B; A will lose more corpuscles than it will gain and so will become positively, while B will be negatively, electrified. This distribution of electricity will tend to diminish the flow of corpuscles from A and increase that from B, and the charges of electricity will accumulate until they have made the two flows equal, when things will be in a steady state. This accumulation of positive electricity on A and of negative on B will form an "electric double layer" between the coatings of which there is a finite potential difference which is a measure of the Peltier effect at the junction of the metals. Similarly, if the flow $\frac{1}{6} n v$ depends upon the temperature of the metal, the transport of corpuscles through each section of an unequally heated conductor will vary, and the state of the conductor cannot be steady: the difference in the amount flowing through different sections will produce an accumulation of electricity along the conductor; this will produce an electric force which, by increasing the flow where it was small and diminishing it where it was large, will make the flow uniform throughout the conductor. These forces represent the Thomson effect.

In the second theory, where the corpuscles are supposed to start from one electric doublet and come to rest on another, there is a movement of corpuscles throughout the body, and we easily see that, with the notation of page 88, the number of corpuscles which pass in one second, in one direction through unit area, is

$$\frac{1}{6} n p b.$$

Hence, as before, if two metals A and B are in contact and if $n p b$ for A is not the same as for B, one metal will gain and the other lose electricity: thus there will be an accumulation of electricity at the junction producing an electric field; this field will increase the flow in one metal and retard it in the other until the two flows are equal.

The manner in which the electric force affects the flow of the corpuscles is essentially different in the two theories. On the first theory the electric force acts on the free corpuscles, accelerating those in one metal and retarding those in the other, while on the second theory the alteration in the flow is brought about by the action of the electric

THEORY OF ELECTRICAL CONDUCTION.

field on the doublets which are supposed to be dispersed through the metal and not on the free corpuscles. If the axes of these doublets are distributed uniformly in all directions, then the flow of corpuscles produced by the detachment of corpuscles from the doublets will be uniform in all directions. If, however, an electric force which we may suppose to be parallel to the axis of x acts on the metal, it will polarise the distribution of the axes of the doublets and will make more point in the direction of the axis of x than in the opposite direction; thus this electric force will diminish the flow of corpuscles along the positive direction of x while it will increase the flow in the opposite direction. We see that by the application of suitable electric forces we can increase or diminish the flow in any direction. At the junction of two metals the initial inequality in the flow of corpuscles across the junction will cause an accumulation of electricity, and this will go on until the forces due to this electrification have made the flows in the two metals equal to each other; it is these forces which give rise to the Peltier effect, while the Thomson effect is represented by the forces which are required to make the flow of corpuscles uniform at all points in an unequally heated conductor.

On the Hall Effect and the Effect of a Magnetic Field on Electrical Resistance.

The Hall effect, on the second theory of metallic conduction, originates in the action of the magnetic field on the distribution of the axes of the doublet which are supposed to exist in the metal, while on the first theory it arose from the action of the magnetic field on the corpuscles.

To see how it arises on the second theory, let us suppose that AB is a doublet and that an electric force parallel to the axis of x acts upon it; this electric field will give rise to a couple tending to bring the axis of the doublet in line with the force. If the motion of the doublet takes place in a magnetic field, then as soon as the doublet begins to move the moving positive and negative charges at its ends will be acted upon by forces which are at right angles to the magnetic force and at right angles also to the direction of motion of the charges. If the doublet were turning about a point midway between the charges, the velocity of the negative charge would be equal and opposite to that of the positive, so that the same force would be exerted by the magnetic field on the two charges, and the joint effect of the two forces on the doublet would be to pull it bodily in some direction without deflecting the axis; in this case there would not be any Hall effect. Suppose, however, that the doublet were not turning about the point midway between the charges, so

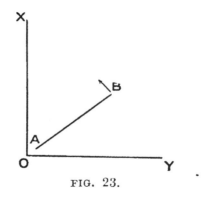

FIG. 23.

that the velocities of these charges were not equal and opposite, then the force due to the magnetic field on the one would not be equal to that on the other; the forces would have a finite moment about an axis through the point about which the doublet is turning, and there would be a couple tending to deflect the axis of the doublet. If the couples arising in this way all tended to twist the axes of the doublets in one direction, there would be an excess of the axes pointing in this direction over those pointing in the opposite, and therefore a current in that direction; thus the magnetic force might give rise to side currents analogous to those which constitute the Hall effect.

To follow this effect into greater detail, let us consider what will happen when the point about which the doublet

AB turns coincides with one of the charges, say, the negative charge A; let the magnetic force act in the direction OY, the electric force along OX (Fig. 23). Consider a doublet AB originally in the plane XOY, then, under the action of the electric force, AB will begin to approach OX, but as the positive charge moves in the magnetic field it will be acted on by a downward force, making B dip below the plane XOY, and thus making the negative end of the doublet be above the positive end. This would have the effect of making more doublets have their negative ends above the positive than below it; there would thus be a vertically downward current of electricity, *i.e.*, a current at right angles to both the magnetic and electric forces, *i.e.*, a current in the direction of the Hall effect. If the positive end of the doublet had been fixed instead of the negative end, the couple tending to twist the axis of the doublet would be reversed, and there would be a majority of doublets having their positive ends above the negative, *i.e.*, there would be a current of electricity vertically *upwards* instead of downwards. This would correspond to a Hall effect of the opposite sign to the preceding. The sign of the Hall effect would depend upon whether the positive or the negative end of the doublet moves the faster when the doublet is deflected by the electric force. The bias which the axis of the doublet experiences in consequence of the magnetic force makes the average angle made by the axis of the doublet with the direction of the electric force greater than it would be if the magnetic force were absent, just as the average angle made with the vertical by a pendulum having for a bob a gyroscope in rapid rotation is greater than that for the pendulum started from the same position with the bob at rest. This increase in the angle between the direction of the electric force and the axes of the doublets means that the polarisation and therefore the electric current is less than it would be if the magnetic force were absent, or, as we may express it, the resistance of the conductor is increased by the magnetic force.

102 THE CORPUSCULAR THEORY OF MATTER.

It will be noticed that the expressions we have found for the electrical and thermal conductivities, the radiation, and the other electrical effects do not involve the mass of the carrier, so that the results would hold if the carriers were bodies having a much greater mass than that of a corpuscle.

CHAPTER VI.

THE ARRANGEMENT OF CORPUSCLES IN THE ATOM.

We have seen that corpuscles are always of the same kind whatever may be the nature of the substance from which they originate; this, in conjunction with the fact that their mass is much smaller than that of any known atom, suggests that they are a constituent of all atoms; that, in short, corpuscles are an essential part of the structure of the atoms of the different elements. This consideration makes it important to consider the ways in which groups of corpuscles can arrange themselves so as to be in equilibrium. Since the corpuscles are all negatively electrified, they repel each other, and thus, unless there is some force tending to hold them together, no group in which the distances between the corpuscles is finite can be in equilibrium. As the atoms of the elements in their normal states are electrically neutral, the negative electricity on the corpuscles they contain must be balanced by an equivalent amount of positive electricity; the atoms must, along with the corpuscles, contain positive electricity. The form in which this positive electricity occurs in the atom is at present a matter about which we have very little information. No positively electrified body has yet been found having a mass less than that of an atom of hydrogen. All the positively electrified systems in gases at low pressures seem to be atoms which, neutral in their normal state, have become positively charged by losing a corpuscle. In default of exact knowledge of the nature of the way in which positive electricity occurs in the atom, we shall consider a case in which the positive electricity is distributed in the way most amenable to mathematical calculation, *i.e.*, when it occurs as a sphere of uniform density, throughout which the corpuscles are distributed. The positive electricity attracts the corpuscles to the centre of the sphere, while their mutual repulsion drives them away from it; when in equilibrium they will be distributed in such a way that the attraction of the positive electrification is balanced by the repulsion of the other corpuscles.

Let us now consider the problem as to how 1...2...3... n corpuscles would arrange themselves if placed in a sphere filled with positive electricity of uniform density, the total negative charge on the corpuscles being equivalent to the positive charge in the sphere.

When there is only one corpuscle the solution is very simple: the corpuscle will evidently go to the centre of the sphere. The potential energy possessed by the different arrangements is a quantity of considerable importance in the theory of the subject. We shall call Q the amount of work required to remove each portion of electricity to an infinite distance from its nearest neighbour; thus in the case of the single corpuscle we should have to do work to drag the corpuscle out of the sphere and then carry it away to an infinite distance from it; when we have done this we should be left with the sphere of positive electricity, the various parts of which would repel each other; if we let these parts recede from each other until they were infinitely remote we should gain work. The difference between the work spent in removing the negative from the positive and that gained by allowing the positive to scatter is Q the amount of work required to separate completely the electrical charges. When there is only one corpuscle we can easily show that $Q = \dfrac{9}{10}\dfrac{e^2}{a}$, where e is the charge on a corpuscle measured in electrostatic units and a is the radius of the sphere.

When there are two corpuscles inside a sphere of positive electricity they will, when in equilibrium, be situated at two points A and B, in a straight line with O the centre of the sphere and such that $OA = OB = \dfrac{a}{2}$, where a is the radius of the sphere. We can easily show that in this position

ARRANGEMENT OF CORPUSCLES IN ATOM. 105

the repulsion between A and B is just balanced by the attraction of the positive electricity and also that the equilibrium is stable. We may point out that AB the distance between the corpuscles is equal to the radius of the sphere of positive electrification. In this case we can show that $Q = \frac{21}{10}\frac{e^2}{a}$.

Thus if the radius of the sphere of positive electrification remained constant, Q for a system containing two corpuscles in a single sphere would be greater than Q for the arrangement in which each corpuscle is placed in a sphere of positive electrification of its own, for in the latter case we have seen that $Q = 2 \times \frac{9}{10}\frac{e^2}{a}$ and this is less than $\frac{21}{10}\frac{e^2}{a}$. Thus the arrangement with the two corpuscles inside one sphere is more stable than that where there are two spheres with a single corpuscle inside each: thus if we had a number of single corpuscles each inside its own sphere, they would not be so stable as if they were to coagulate and form systems each containing more than one corpuscle. There would therefore be a tendency for a large number of systems containing single corpuscles to form more complex systems. This result depends upon the assumption that the size of the sphere of positive electrification for the system containing two corpuscles is the same as that of the sphere containing only one corpuscle. If we had assumed that when two systems unite the volume of the sphere of positive electricity for the combined system is the sum of the volumes of the individual systems, then a for the combined system would be $2\frac{1}{3}$ or 1·25 times a for the single system. Taking this into account, we find that Q for the combined system is less than the sum of the values of Q for the individual system; in this case the system containing two corpuscles would not be so stable as two systems each containing one corpuscle, so that the tendency now would be towards dissociation rather than association.

Three corpuscles inside a single sphere will be in stable equilibrium when at the corners of an equilateral triangle

106   THE CORPUSCULAR THEORY OF MATTER.

whose centre is at the centre of the sphere and whose side is equal in length to the radius of that sphere; thus for three as for two corpuscles the equilibrium position is determined by the condition that the distance between two corpuscles is equal to the radius of the sphere of positive electrification.

For the case of three corpuscles $Q = \frac{26}{10}\frac{e^2}{a}$, and thus again we see that if the radius of the sphere of positive electricity is invariable, the arrangement with three corpuscles inside one sphere is more stable than three single corpuscles each inside its own sphere, or than one corpuscle inside one sphere and two corpuscles inside another sphere; thus again the tendency would be towards aggregation. If, however, the positive electricity instead of being invariable in size were invariable in density, we see that the tendency would be for the complex system to dissociate into the simpler ones.

Four corpuscles if at rest cannot be in equilibrium when in one plane, although the co-planar arrangement is possible and stable when the four are in rapid rotation. When there is no rotation the corpuscles, when in stable equilibrium, are arranged at the corners of a regular tetrahedron whose centre is at the centre of the sphere of positive electrification and whose side is equal to the radius of that sphere; thus we again have the result that the distance between the corpuscles is equal to the radius of the positive sphere.

For four corpuscles $Q = \frac{e^2}{a}\frac{54}{10}$. We see that the values of Q per corpuscle are for the arrangements of 1, 2, 3, 4 corpuscles in the proportion of 6 : 7 : 8 : 9 if the radius of the positive sphere is invariable.

Six corpuscles will be in stable equilibrium at the corners of a regular octahedron, but it can be shown that the equilibrium of eight corpuscles at the corners of a cube is unstable. The general problem of finding how n corpuscles will distribute themselves inside the sphere is very complicated, and I have not succeeded in solving it; we can, however, solve the special case where the corpuscles are

ARRANGEMENT OF CORPUSCLES IN ATOM. 107

confined to a plane passing through the centre of the sphere, and from the results obtained from this solution we may infer some of the properties of the more general distribution. The analytical solution of the problem when the motion of the corpuscles is confined to one plane is given in a paper by the author in the *Philosophical Magazine* for March, 1904; we shall refer to that paper for the analysis and quote here only the results.

If we have n corpuscles arranged at the corners of a regular polygon with n sides with its centre at the centre of the sphere of positive electrification, each corpuscle being thus at the same distance r from the centre of this sphere, we can find a value of r, so that the repulsion exerted by the $(n-1)$ corpuscles on the remaining corpuscle is equal to the attraction of the positive electricity on that corpuscle; the ring of corpuscles would then be in equilibrium. But it is shown in the paper referred to that if n is greater than 5 the equilibrium is unstable and so cannot exist; thus 5 is the greatest number of corpuscles which can be in equilibrium as a single ring. It is shown, however, that we *can* have a ring containing more than five corpuscles in equilibrium if there are other corpuscles inside the ring. Thus, though a ring of six corpuscles at the corners of a regular hexagon is unstable by itself, it becomes stable when there is another corpuscle placed at the centre of the hexagon, and rings of seven and eight corpuscles are also made stable by placing one corpuscle inside them. To make a ring of nine corpuscles stable, however, we must have two corpuscles inside it, and the number of corpuscles required inside a ring to keep it stable increases very rapidly with the number of corpuscles in the ring. This is shown by the following Table, where n represents the number of corpuscles in the ring and i the number of corpuscles which must be placed inside the ring to keep it in stable equilibrium :—

n.	5.	6.	7.	8.	9.	10.	12.	13.	15.	20.	30.	40.
i.	0.	1.	1.	1.	2.	3.	8	10.	15.	39.	101.	232.

When n is large i is proportional to n^3. We thus see that in the case when the corpuscles are confined to one plane they will arrange themselves in a series of concentric rings. When we have determined the relation between n and i, i.e., have found that $i = f(n)$ where f is a known function, the problem of finding the configuration of N corpuscles when in stable equilibrium admits of a very simple solution. When the number of rings is as small as possible the number of corpuscles in each ring will be as large as possible. If n_1 is the number in the outer ring then there will be $N - n_1$ inside it, and if these are just sufficient to keep the outer ring in stable equilibrium $N - n_1 = f(n_1)$,

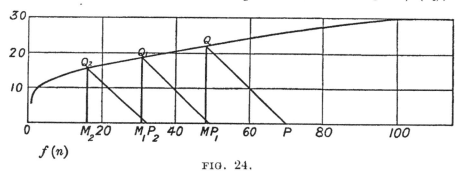

FIG. 24.

the solution of this equation will give us n_1. To find n_2, the number in the next ring, we evidently have the equation—

$$N - n_1 - n_2 = f(n_2),$$

while n_3, the number in the next ring, is given by the equation—

$$N - n_1 - n_2 = f(n_3),$$

and so on.

These equations can be solved very rapidly by a graphical method. Draw the graph whose abscissa $= f(n)$ and whose ordinate is n; the values of $f(n)$ for a series of values of n are given on page 107; from these values the curve (Fig. 24) has been constructed.

To find how a number of corpuscles equal to N will arrange themselves, measure off along the axis of abscissae a distance from O equal to N. Let OP be this distance; through P draw the straight line PQ inclined at an angle

ARRANGEMENT OF CORPUSCLES IN ATOM.

of 135° to the horizontal axis, intersecting the curve in Q, draw the ordinate QM, then the integral part of QM will be the value of n_1, the number of corpuscles in the outside ring. For evidently

$$OM = f(QM)$$

and $OM = OP - PM$, and since PQ is inclined at 45° to the axis, $QM = PM$; hence

$$OP - QM = f(QM).$$

comparing this with the equation $N - n_1 = f(n_1)$, we see that the integral part of QM is equal to n_1.

To get the value of n_2, the number of the corpuscles in the second ring, we mark off the abscissa $OP_1 = N - n_1$; if QM is an integer P_1 will coincide with M; from P_1 draw $P_1 Q_1$ parallel to PQ, cutting the curve in Q_1; if $Q_1 M_1$ is the ordinate at Q_1, the integral part of $Q_1 M_1$ will be the value of n_2. To get n_3 mark off $OP_2 = N - n_1 - n_2$ and draw $P_2 Q_2$ parallel to PQ; the integral part of $Q_2 M_2$ will be the value n_3. In this way we can, in a very short time, find the numbers of corpuscles in the various rings.

The following Table giving the various rings for corpuscles ranging in number from 1 to 100 has been calculated in this way; the first row contains the numbers for which there is only one ring, the second those with two rings, the third those with three, and so on:—

NUMBERS OF CORPUSCLES IN ORDER.

```
 1  2  3  4  5

 5  6  7  8  8  8  9 10 10 10 11
 1  1  1  1  2  3  3  3  4  5  5

11 11 11 12 12 12 13 13 13 13 13 14 14 15 15
 5  6  7  7  8  8  8  8  9 10 10 10 10 10 11
 1  1  1  1  1  2  3  3  3  3  4  4  5  5  5

15 15 15 16 16 16 16 16 16 17 17 17 17 17 17 17
11 11 11 11 12 12 12 13 13 13 13 13 13 14 14 15
 5  6  7  7  7  8  8  8  9  9 10 10 10 10 10 11
 1  1  1  1  1  1  2  2  3  3  3  3  4  4  5  5  5

17 18 18 18 18 18 19 19 19 19 20 20 20 20 20 20 20 20 21 21
15 15 15 16 16 16 16 16 16 16 16 16 17 17 17 17 17 17 17 17
11 11 11 11 11 12 12 12 12 13 13 13 13 13 13 14 14 15 15 15
 5  5  6  7  7  7  7  8  8  8  8  9  9 10 10 10 10 10 10 11
 1  1  1  1  1  1  1  2  2  2  3  3  3  3  4  4  5  5  5  5
```

110   THE CORPUSCULAR THEORY OF MATTER.

NUMBER OF CORPUSCLES IN ORDER—*continued*.

21	21	21	21	21	21	21	21	22	22	22	22	22	22	22	22	23	23	23	23	23	23	23	24
17	18	18	18	18	18	19	19	19	19	19	20	20	20	20	20	20	20	20	20	20	21	21	21
15	15	15	15	16	16	16	16	16	16	16	16	16	17	17	17	17	17	17	17	17	17	17	17
11	11	11	11	11	12	12	12	12	12	13	13	13	13	13	13	13	13	14	14	15	15	15	15
5	5	6	7	7	7	7	8	8	8	8	8	8	9	9	10	10	10	10	10	10	10	11	11
1	1	1	1	1	1	1	1	1	2	2	2	3	3	3	3	3	4	4	5	5	5	5	5

24	24	24	24	24	24	24
21	21	21	21	21	21	21
17	18	18	18	18	18	19
15	15	15	15	16	16	16
11	11	11	11	11	12	12
5	5	6	7	7	7	7
1	1	1	1	1	1	1

We can investigate the equilibrium of corpuscles in one plane by experiment as well as by analysis, using a method introduced for a different purpose by an American physicist, Professor Mayer. The problem of the arrangement of the corpuscles is to find how a number of bodies which repel each other with forces inversely proportional to the square of the distance between them will arrange themselves when under the action of an attractive force tending to drag them to a fixed point. For the experimental method the corpuscles are replaced by magnetised needles pushed through cork discs and floating on water. Care should be taken that the needles are equally magnetised. These needles, having their poles all pointing in the same way, repel each other like the corpuscles. The attractive force is produced by a large magnet placed above the surface of the water, the lower pole of this magnet being of the opposite sign to that of the upper poles of the floating magnets. The component along the surface of the water of the force due to this magnet is directed to the point on the surface vertically below the pole of the magnet, and is approximately proportional to the distance from this point. The forces acting on the magnets are thus analogous to those acting on the corpuscles.

If we throw needle after needle into the water we shall find that they will arrange themselves in definite patterns, three needles at the corners of a triangle, four at the corners of a square, five at the corners of a pentagon; when, however,

ARRANGEMENT OF CORPUSCLES IN ATOM. 111

we throw in a sixth needle this sequence is broken; the six needles do not arrange themselves at the corners of a hexagon, but five go to the corners of a pentagon and one goes to the middle. When we throw in a seventh needle we get a ring of six with one at the centre; thus a ring of six, though unstable when hollow, becomes stable as soon as one is put in the inside. This is an example of a fundamental principle in the stable configurations of corpuscles; the

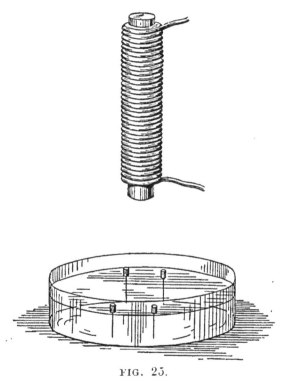

FIG. 25.

structure must be substantial; we cannot have a great display of corpuscles on the outside and nothing in the inside. If, however, we have a good foundation of corpuscles—if, for example, we tie a considerable number of needles together for the inside—we can have a ring containing a large number of corpuscles in stable equilibrium around it, although five is the greatest number of corpuscles that can be in equilibrium in a hollow ring. By the aid of these floating magnets we can illustrate the configurations for considerable numbers of corpuscles, and verify the Table previously given.

Another method, due to Professor R. W. Wood, is to replace the magnets floating on water by iron spheres floating on mercury; these spheres get magnetised by induction by the large magnet placed above them and repel each other—though in this case the repulsive force does not vary inversely as the square of the distance—while they are attracted by the external magnet; the iron spheres arrange themselves in patterns analogous to those formed by the magnets. Dr. Monckman used, instead of magnets, elongated conductors floating vertically in water; these were electrified by induction by a charged body held above the surface of the water; the conductors, being similarly electrified, repelled each other and were attracted towards the electrified body; under these forces they formed patterns similar to those formed by the floating magnets.

We see from this experimental illustration, as well as by the analytical investigation, that a number of corpuscles will, if confined to one plane, arrange themselves in a series of rings, the number of corpuscles in the ring increasing as the radius of the ring increases.

If we refer to the arrangements of the different numbers of corpuscles given on page 109, we see that the numbers which come in the same vertical columns are arranged in patterns which have much in common, for each arrangement is obtained by adding another storey to the one above it. Thus, to take the first column, we have the pattern 5, 1, the one below it is 11, 5, 1; the one below this 15, 11, 5, 1; the one below this 17, 15, 11, 5, 1; then 21, 17, 15, 11, 5, 1; and then 24, 21, 17, 15, 11, 5, 1. We should expect the properties of the atoms formed of such arrangements of corpuscles to have many points of resemblance. Take, for example, the vibrations of the corpuscles; these may be divided into two sets. The first set consists of those arising from the rotation of the corpuscles around their orbits. If all the corpuscles in an atom have the same angular velocity, the frequency of the vibrations produced by the rotation of the ring of corpuscles is proportional to the number of corpuscles in the ring; and thus in the spectrum of each of the elements corresponding to the arrangements of corpuscles found in a vertical column in the Table, there would be a series of lines whose frequencies would be in a constant ratio to each other, this ratio being the ratio of the numbers of corpuscles in the various rings.

The second set of vibrations are those corresponding to the displacement of a ring from its circular shape. If the distance of a corpuscle from the nearest member in its own ring is small compared with its distance from its nearest neighbour on another ring, the effect of the outer ring will only "disturb" the vibrations of the ring without altering their fundamental character. Thus we should expect the various elements in a vertical column to give corresponding groups of associated lines. We might, in short, expect the various elements corresponding to the arrangements of the corpuscles contained in the same vertical column, to have many properties, chemical as well as physical, in common. If we suppose that the atomic weight of an element is proportional to the number of corpuscles contained in its atom,—and we shall give later on evidence in favour of this view,—we may regard the similarity in properties of these arrangements of corpuscles in the same vertical column as similar to a very striking property of the chemical elements, *i.e.*, the property expressed by the periodic law. We know that if we arrange the elements in the order of their atomic weights, then as we proceed to consider the elements in this order, we come across an element—say lithium—with a certain property; we go on, and after passing many elements which do not resemble lithium, we come to another, sodium, having many properties in common with lithium; then, as we go on we lose these properties for a time, coming across them again when we arrive at potassium, and so on. We find here just the same recurrence of properties at considerable intervals that we should get if the atoms contained numbers of corpuscles proportional to their atomic weight. Consider a series of atoms, such that the atom of the pth member is formed from that of the $(p-1)$th by the addition of a single ring, *i.e.*, is a compound, so to speak, of the $(p-1)$th atom with a fresh ring. Such a series would belong to elements which are in the same group according to the periodic law, *i.e.*, these elements form a series which, if arranged according to Mendeleef's table would all be in the same vertical column.

The properties of these configurations of corpuscles have further analogies with the properties of real atoms. To illustrate this let us consider the properties of all the configurations of corpuscles which have 20 corpuscles in the outside ring. The smallest number of corpuscles which has an outer ring of 20 is 59; in this case the number of corpuscles inside the ring is only just sufficient to make the outer ring stable, this ring will therefore be on the verge of instability, and when the corpuscles in the ring are displaced the forces of restitution urging them back to their original positions will be small. Thus, when this ring is subject to disturbances from an external source, a corpuscle will easily be detached from it, and the group will, by losing a negatively electrified corpuscle, acquire a charge of positive electricity; the group will thus resemble the atom of a strongly electropositive element. When we pass from 59 to 60 corpuscles the outer ring is more stable because there are more corpuscles inside it; the corresponding atom will, therefore, not be so strongly electropositive as that containing only 59 corpuscles. The addition of each successive corpuscle will make it more difficult to detach corpuscles from the outer ring, and will, therefore, make the corresponding atom less electro-positive. The increase in the stability of the ring, and consequently in the electro-negative character of the corresponding atom, will go on increasing until we have as many as 67 corpuscles in the group, when the stability of the outer ring will be a maximum. A great change in the properties of the group will occur when the number of corpuscles increases from

ARRANGEMENT OF CORPUSCLES IN ATOM.

67 to 68, for with 68 corpuscles the number in the outer ring is 21; these 21 corpuscles are, however, only just stable, and, like the outer ring of 20 in the arrangement of 59 corpuscles, would readily lose a corpuscle. The atom corresponding to this arrangement would, therefore, be strongly electro-positive.

The properties of the groups of 59 and 67 corpuscles, which are respectively at the beginning and end of the series which has 20 corpuscles in the outer ring, deserve especial consideration. The arrangement of corpuscles in the group of 59, though near the verge of instability and therefore very liable to lose a negative corpuscle and thus acquire a positive charge, would not be able to retain this charge. For when it had lost a corpuscle the 58 corpuscles left would arrange themselves in the grouping corresponding to 58 corpuscles, which is the last to have an outer ring of 19 corpuscles; this ring is therefore exceedingly stable, so that no further corpuscles would escape from it, while the positive charge on the system due to the escape of the 59th corpuscle would attract the surrounding corpuscles. Thus this arrangement could not remain permanently charged with positive electricity, for as soon as one corpuscle escaped it would be replaced by another. If, however, corpuscles were shot into the arrangement of 59 from outside, each additional corpuscle would increase the stability of the system until the number reached 67; the arrangement corresponding to 68 would be very unstable, so when this number was reached the system would lose corpuscles. Thus a charge of 8 units of negative electricity could be forced into this group, which would correspond, therefore, to an atom with a valency 0 for a positive charge, and a valency 8 for a negative one.

Let us now consider the properties of the group of 67 corpuscles. The outer ring of this would be very stable, but if an additional corpuscle were added to the group the 68 corpuscles would arrange themselves with a ring of 21 on the outside, as 68 is the smallest number of corpuscles with an outer ring of 21; the ring is very unstable, and easily loses the corpuscle it has gained, thus the arrangement could not permanently be negatively charged—it would act like the atom of an element of no electro-negative valency. On the other hand, the arrangement would be stable if one, two, three up to eight corpuscles were abstracted from it, although from the firmness with which they are held this detachment of corpuscles would be difficult; as each corpuscle abstracted leaves the arrangement with a positive charge, the work required to remove the successive corpuscles would tend to increase. This tendency would be to some extent compensated by the diminishing stability of the arrangements 66, 65, 64,...59; but when once 59 is reached, not only have we to overcome the positive charge, but also the great stability of the arrangement of 58 corpuscles, so that eight would be the greatest number of corpuscles we could hope to remove from the group; thus the atom represented by this group would have an electro-positive valency represented by 8, while its electro-negative valency is zero.

Let us now consider the group containing 60 corpuscles. This will be the most electro-positive of the series; it can, however, only retain permanently a positive charge of one unit of electricity, corresponding to the removal of one corpuscle, for when it has lost two corpuscles we should have the group 58 as we had when we removed one corpuscle from the group 59; and in the present case the group would be more likely to attract a corpuscle than when we started from 59 instead of 60, for it would have a charge of two positive units instead of one. Thus the atom represented by the group of 60 would have an electro-positive valency of one. If we force additional corpuscles into the group so that the corpuscles increase to 61, 62, 63, 67, the arrangements become more and more stable; when, however, we get to 68 we have an arrangement which is nearly unstable and which will readily give off corpuscles. Thus seven is the greatest number of corpuscles we could hope to force into this group, so that the atom represented by it would have an electro-negative

ARRANGEMENT OF CORPUSCLES IN ATOM.

valency of seven. We have seen that the electro-positive valency is one.

The group of 66 corpuscles would be the most electro-negative of the series, but would only be able to retain a charge of one unit, for if it acquired two units the group would consist of 68 corpuscles, an arrangement which, as we have seen, rapidly loses its corpuscles. The atom corresponding to the group 66 will thus have an electro-negative valency of one. We see, too, that seven corpuscles could be extracted from the group without destroying its stability; thus the atom corresponding to this group would have an electro-positive valency seven.

The group of 61 corpuscles would not part with its corpuscles so readily as the group of 60, but on the other hand it could afford to lose two, as it is not reduced to 58 corpuscles until it has lost a third corpuscle, and 58 is the number when the tendency to attract and retain corpuscles would suddenly rise; thus the atom corresponding to the group 61 would have an electro-positive valency of 2. In the same way as before we see that it could find room for 6 corpuscles, so that the corresponding atom might have an electro-negative valency of 6. In a similar way we see that the group of 62 would correspond to an electro-negative atom with an electro-negative valency of 3 and an electro-positive valency of 5. The group 63 is an atom with an electro-negative valency of 4 and an electro-positive valency of 4. Thus, tabulating our results, we have the following properties of the series of atoms corresponding to the groups containing from 59 to 67 corpuscles:—

No. of corpuscles	59	60	61	62	63	64	65	66	67
Valency	$+0$	$+1$	$+2$	$+3$	$+4$	-3	-2	-1	-0
	-8	-7	-6	-5	-4	$+5$	$+6$	$+7$	$+8$

Electro-positive.   Electro-negative.

This sequence of properties is very like that observed in the case of the atoms of the elements.

Thus we have the series of elements—

He. Li. Be. B. C. N. O. F. Ne.
Ne. Na. Mg. Al. Si. P. S. Cl. Arg.

The first and last element in each of these series has no valency, the second is a monovalent electro-positive element, the last but one a monovalent electro-negative element, the third is a divalent electro-positive element, the last but two a divalent electro-negative element, and so on.

In our Table we have assigned two valencies to the element according as it acts as an electro-positive or an electro-negative element, and we notice that the sum of these valencies is constant and equal to 8. It is interesting to find that Abegg,[1] from purely chemical considerations, shows that the valency of an element is very different when it acts as the electro-positive constituent of a compound from its valency when it is the electro-negative constituent. Thus chlorine has the valency 1 in a compound like HCl, in which it is the electro-negative constituent, but has much higher valencies when combined with very electro-negative elements such as oxygen. Iodine, too, is another striking instance; it is monovalent when combined with electro-positive elements like the metal, but has a much higher valency when combined with more electro-negative elements as in the compound $I\ Cl_5$. The view that the same element is sometimes the positive constituent and in other combinations the negative constituent of a compound has received further confirmation recently by some remarkable experiments made by Walden.

The sum of the positive and negative valencies would depend upon the number of corpuscles assumed to be in the outer ring. If we take the number in the outer ring to be 20 the sum of the positive and negative valencies is 8; this happens to agree with the number usually assigned to this sum by chemists; this agreement with the results given by the model atom is, however, quite accidental.

[1] "Zeitschrift für Anorganische Chemie," 39, p. 330, 1904; "Zeitschrift für Physikalische Chemie," 43, p. 385, 1903.

ARRANGEMENT OF CORPUSCLES IN ATOM.

It may not be out of place here to again emphasise the statement that the special arrangement of corpuscles in which they are supposed to be confined to one plane, and in which the positive electricity attracts them with a force proportional to their distance from a fixed point, has been chosen because it is the one most amenable to mathematical treatment. My object has been to show that stable arrangements of corpuscles would have many properties in common with real atoms, and I have attempted to illustrate these properties by considering a special case chosen solely on the ground of simplicity. The number of corpuscles corresponding to any particular property would doubtless be different if we took a three- instead of a two- dimensional distribution of corpuscles, or if instead of supposing the attractive force exerted by the positive electricity to vary directly as the distance from a fixed point we assumed that the density of the positive electricity inside the sphere was not uniform, in which case the attraction would follow a much more complicated law.

The two-fold valency would be a property of the atom whatever its structure, provided that, as in the special case of corpuscles confined to one plane, there is a great change of stability in passing through certain groups of corpuscles, the number of corpuscles in these critical groups being, say, N_1, N_2, N_3. . . . The work required to add a corpuscle to, or take one away from, a group of corpuscles would be abnormally great when the change in the number of corpuscles involved the passage through or into one of these critical numbers; thus these critical numbers may be regarded as barriers which cannot easily be passed. As an atom containing $N_2 + n$ corpuscles could lose n corpuscles and gain $N_3 - (N_2 + n)$ without crossing one of these barriers, such an atom would have a maximum positive valency n and a maximum negative valency $N_3 - (N_2 + n)$.

We may also look at the question from the following point of view: we may express the tendency of a group of corpuscles to shed a corpuscle as arising from the equivalent of a corpuscular pressure in the atom, and we may express the preceding result by saying that when the number of corpuscles increases through one of the values $N_1, N_2, N_3 \ldots$, say N_1, the corpuscular pressure abruptly increases, and then falls gradually as the number of corpuscles increases to N_2, when again the pressure abruptly increases. Thus for a group of corpuscles intermediate in number between N_1 and N_2 we could go on adding corpuscles without increasing the corpuscular pressure (though of course we should increase the repulsion arising from the negative charge on these corpuscles) until we reached N_2, but since at N_2 the corpuscular pressure rapidly increases, we could not without great difficulty increase the number of corpuscles to $N_2 + 1$. Again, we could take away corpuscles from the original group without diminishing the corpuscular pressure until the number of corpuscles is reduced to N_1. Since the corpuscular pressure abruptly falls at this point it would be difficult to extract another corpuscle from the group. Thus if N the number of corpuscles in this group $= N_1 + n$, the maximum number of corpuscles we could extract would be n, *i.e.*, the maximum positive valency would be n, while the greatest number we could add would be $N_2 - (N_1 + n)$, and this would be the maximum negative valency.

FORCES BETWEEN THE ATOMS. CHEMICAL COMBINATION.

A very important and interesting subject of investigation is the nature of the forces that would be exerted between groups of corpuscles and its application to the theory of chemical combination.

We shall begin by considering the forces between two groups in some simple cases. Let us begin with the simplest of all when we have a single corpuscle at the centre of a sphere of positive electrification. Let us take two such systems equal in all respects, then as long as one is wholly outside the other there will be neither attraction nor repulsion between the systems; when, however, the spheres cut, as in Fig. 26, the systems will attract each other. To see this, consider the action of the system A on B; there

ARRANGEMENT OF CORPUSCLES IN ATOM. 121

will be no action on that part of B which is outside A, while the action on the part of the positive electricity of B which is inside A will be an attraction towards the centre of A, for inside a sphere the force due to the negative corpuscle at the centre is greater than that due to the positive electricity. The corpuscles will remain at the centres of their respective spheres until they come so close together that the centre of one sphere lies inside the other sphere; when this stage is reached the corpuscles begin to be displaced and are pushed apart so as to be outside the line joining the two centres. In this case there is no difference in the electrification of the spheres; we cannot say that one is positively the other negatively electrified; and if the spheres were separated

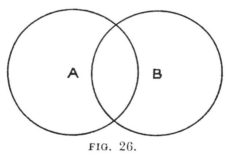

FIG. 26.

after having been together they would each be neutral, the positive electricity in each sphere being balanced by the negative charge at the centre. We thus see that it is possible to have forces electrical in their origin binding the two systems together without a resultant charge on either system. If, however, the spheres were of very different size, then if they were brought close enough together the two corpuscles would go inside one sphere and would remain so after the spheres were pulled apart; thus one sphere would be positively, the other negatively, electrified. Lord Kelvin has shown that it is the smaller sphere that acquires the additional corpuscle, and has proved that when two spheres whose radii are in the proportion of 3 to 1 are gradually brought together, the corpuscle which originally was at the centre of the larger sphere will be transferred to the inside of the smaller when the distance between the centres of the two spheres is reduced to between 2·6 and 2·7 times the radius of the smaller sphere. With systems containing only one corpuscle the only way in which one system can differ from the other is in the size of the sphere of positive electrification. The preceding result is a special case of the general principle that a transference of corpuscles from one group of corpuscles to another different group may occur when the two groups are brought close together. The general nature of this effect may be understood from the following considerations. If we have two groups of corpuscles A and B such that the work required to detach a corpuscle from A is less than that required to detach one from B, then when A and B are brought close together there will be a tendency for a corpuscle to go from A into B, and therefore for A to get positively, B negatively, electrified. The system A corresponds—to take the case considered on page 117—to the groups in the earlier part of the series from 59 to 67, the system B to the groups in the later part. We may represent this effect in a way which is easily handled, by saying that inside the group of corpuscles, or atom, there is a certain corpuscular pressure, and that when two atoms are brought near together the corpuscles tend to pass from the atom where the corpuscular pressure is high to one where it is low. This corpuscular pressure, by which we represent the electric forces inside the atom, is high when the work required to detach a corpuscle from the atom is small, low when this work is large. Thus in our example the corpuscular pressure is high when the number of corpuscles inside the outer ring is only just sufficient to make that ring stable, while it is low when the number of corpuscles inside is considerably greater than the minimum number required for equilibrium, $i.e.$, the pressure is high in the electropositive elements, low in the electro-negative ones. We see, too, that the positive valency of an electro-positive element is the greatest number of corpuscles it can lose before the corpuscular pressure suffers a great diminution.

ARRANGEMENT OF CORPUSCLES IN ATOM. 123

To take an illustration: in the group of 60 corpuscles we say that the corpuscular pressure is high since there is only one more corpuscle inside the outer ring than the minimum number required to make that ring stable; if, however, two corpuscles were to leave the system, the number would be reduced to 58, and the group 58 has an outer ring of 19 with the maximum number of corpuscles inside it; the system has thus great stability, and would thus be represented by one with a low corpuscular pressure. The negative valency of an electro-positive element is the greatest number of corpuscles which can be added without producing a sudden increase in the corpuscular pressure. Thus, to take the case quoted before, if eight corpuscles were added to the group of 60 corpuscles, we should get 68 corpuscles; now 68 is the smallest number of corpuscles which has as many as 21 corpuscles in the outer ring, thus there is only the minimum number inside required for stability, so that the corresponding corpuscular pressure is very high, while if we added seven corpuscles to the group of 60 we should have 67 corpuscles, and as this is the largest number of corpuscles which has 20 in the outside ring, there is the maximum number inside the ring; the stability is thus very great, and the corresponding corpuscular pressure low. We thus see that seven is the greatest electro-negative valency for the group 60.

The negative valency of the electro-negative elements, which have atoms in which the corpuscular pressure is low, is the number of corpuscles which can be added without producing a sudden increase in the corpuscular pressure. Thus, to take an example, the atom corresponding to the group of 66 corpuscles has an electro-negative valency of one, for if it received 2 corpuscles the corpuscles would arrange themselves like the group 68, which, as we have seen, has a very high corpuscular pressure.

The electro-positive valency of these elements is the greatest possible number of corpuscles which can be taken from them without producing an abrupt diminution in the corpuscular pressure. Thus, to take the group of 66, if we take away 7 corpuscles we get the arrangement corresponding to 59, which is nearly unstable, and in which, therefore, the corpuscular pressure is very high. If, however, we take away 8 corpuscles, we have only 58 left, and this arrangement is very stable, as it is the largest number of corpuscles which has as few as 19 in the outer ring, so that the corpuscular pressure is very low; thus the corpuscular pressure is diminished abruptly when, having already detached 7, we detach an additional corpuscle; hence we see that the electro-positive valency of the group 60 is 7.

To sum up, if the electro-positive valency of an atom is n, then we can extract n corpuscles without diminishing the corpuscular pressure, but if we detach an additional corpuscle the corpuscular pressure will suddenly fall; if the electro-negative valency of an atom is m, we can add m corpuscles without increasing the corpuscular pressure, but the addition of the $(m + 1)$th will cause a large increase in the pressure.

Let us see now how these considerations apply to the chemical combinations of the different elements. Let us suppose that we have two different atoms A and B close together, and that the element of which A is an atom is more electro-positive than the one of which B is an atom; this on our view means that the corpuscular pressure in the atom A is greater than that in B, so when A and B are put close together a corpuscle will tend to go from A to B, and thus A would get positively, B negatively electrified. The loss of the corpuscle by the electro-positive atom would, if its positive valency were greater than unity, tend to increase the corpuscular pressure in A, while the gain of a corpuscle by B would, unless it were negatively univalent, lower its corpuscular pressure; this effect would tend to make the flow of corpuscles from A to B continue; on the other hand, the positive electrification on A and the negative on B would tend to stop the flow. Let us suppose that the electro-positive valency of A is unity, then if a second corpuscle were to escape from A the corpuscular pressure, as we have seen above, would fall abruptly, so that instead of the

ARRANGEMENT OF CORPUSCLES IN ATOM. 125

pressure gradient being down from A to B it would be in the opposite direction, and the corpuscle would come back; thus A would not lose more than one corpuscle. If the negative valency of B were unity then B would not be in a condition to receive more than one corpuscle, for if it received two the corpuscular pressure would suddenly increase, and corpuscles would tend to leave B instead of coming into it; if, however, B had an electro-negative valency of 2 it could receive another corpuscle without increase in pressure, and though it could not receive this from A, if another atom A' similar to A were brought to it, a corpuscle might flow from A' into B, and thus B would get charged with two units of negative electricity, A and A' each having a positive charge of one unit; thus B might hold by the electrostatic attraction two atoms A and A' in combination. It could not, however, hold a third atom, because if another atom A'', similar to A and A', were brought up to it, and if another corpuscle were to flow into B, the corpuscular pressure in B would experience a large increase, since, as its valency is only two, it cannot receive more than two corpuscles without having its corpuscular pressure largely increased. Thus B can hold two, but not more than two, univalent atoms in combination; if, however, B had been trivalent instead of divalent, it could receive 3 corpuscles without increasing the corpuscular pressure, so that a corpuscle might flow from a third atom A'' into B, and B be able to keep 3 atoms A, A', A'', in combination. It must be noticed, however, that the transference of the corpuscles from the atoms A', A'', which are supposed to come into the neighbourhood of B after A has transferred its corpuscles, will take place under less favourable conditions than the transference of the corpuscle from A, the atom first brought into its neighbourhood. For when A approached B both were supposed uncharged, but after the corpuscle from A has gone into B, B has a negative charge, and the corpuscle going from A' will have to surmount the electrostatic repulsion due to the charge. Again, after A' has discharged its corpuscle, B will be charged with two units of negative electricity, and the corpuscle going from A'' will have to overcome a greater repulsion than that experienced by the corpuscle from A'. Thus we see that in the case of a multivalent atom it may be more difficult to fill up the later valencies than the earlier ones. The existence of "unsaturated" compounds, such as PCl_3, $MnCl_2$, may be taken as an illustration of this point.

Again, this difficulty will produce the most marked effects when the difference of corpuscular pressure tending to drive the corpuscles from one atom to the other is small, *i.e.*, when the elements resemble each other in properties. We should thus expect that the valency of one element towards an element of somewhat similar properties would be less than its valency towards a widely dissimilar element.

The terms electro-negative and electro-positive are only relative, and an element may be electro-positive to one substance and electro-negative to another. It would appear from the preceding considerations that the valency of an element where it is acting as the electro-negative constituent of a compound may be very different from the valency when it is the electro-positive constituent. Thus, to take the group of 60 corpuscles as an example, when it is in combination with a more electro-negative element, *i.e.*, one where the corpuscular pressure is lower, it can, as we have seen, only lose one corpuscle, *i.e.*, its electro-positive valency is one. But if the group 60 were placed near a group G with a still higher corpuscular pressure, so that corpuscles flow into the group 60 instead of flowing out of it, then, since we have seen that the corpuscular pressure of the group 60 will not be suddenly increased by the addition of corpuscles until the number of corpuscles added exceeds 7, we see that the group of 60 might receive as many as 7 corpuscles from such groups as G, and would therefore have a valency 7. There are many compounds which suggest a difference of this kind. Thus iodine appears monovalent in the compound $H\,I$, in which it is the electro-negative element, while it is sexavalent in the compound $I\,F_6$, in which it is probably the positive element.

We see that on these views the valency of an element is not a constant quantity; it depends on whether the element is the electro-positive or electro-negative constituent of the compound, and even when the sign of its charge is the same, on the nature of the element with which it is in combination, an element having a smaller valency when combined with one of similar properties than when in combination with one from which it differs more widely.

In the cases of chemical combination we have considered, we have supposed that there is a transference of corpuscles from one atom to the other, and that the attraction between the positive and negative electrification resulting from this transference helps to bind together the elements in the compound. The case, however, of two equal spheres, each with a single corpuscle at its centre, considered on p. 121, shows that there may be attractions between atoms consisting of groups of corpuscles without any transference of corpuscles, *i.e.*, when neither atom gets charged with electricity. A very important question arises when we consider the combination of two similar atoms in the molecule of an elementary gas: is there or is there not a transference of electricity in this case—*i.e.*, does one atom acquire a charge of positive, the other of negative, electricity? If two similar atoms or groups of corpuscles are brought together, a symmetrical distribution of corpuscles, *i.e.*, one in which there is no transference of corpuscles, will certainly be one of equilibrium. The question, however, is, is the equilibrium stable? We can easily give examples in which the equilibrium of symmetrical arrangements is unstable. Take for example the case of two electrified drops of water placed in a vessel which they very nearly fill, let condensation on the sides of the vessel be prevented, so that the vapour from one drop condenses on the other. There will be equilibrium if the drops are equal, but this equilibrium will be unstable, for if one drop were to differ ever so little in size from the other, the smaller drop would evaporate more rapidly than the larger; thus the big drop would get bigger by condensation, and the little drop smaller. When the little drop got below a certain size the electrical charge would so lower its vapour pressure that it would sink to that of the big drop, and there would be equilibrium, and in this case the equilibrum would be stable, for if the little drop were to get smaller its vapour pressure would diminish so rapidly that water would condense on it and it would grow bigger; while if it got bigger the vapour pressure would increase and the drop would become smaller. Thus

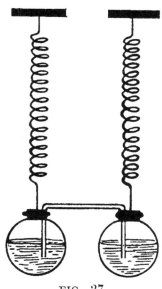

FIG. 27.

two charged drops of water, equal in all respects to begin with, will not remain equal, and the stable configuration will not be two equal drops, but one big and one little drop.

Another example in which the forces concerned bear a somewhat close analogy to those at work in the atom is the following :—

Let us represent an atom in the normal state by a closed glass vessel partly filled with water and suspended from a spring balance. To represent the effect of the influence of an atom on an equal atom near to it, let us suppose that the water in two similar vessels is connected by a syphon, as in

ARRANGEMENT OF CORPUSCLES IN ATOM.

Fig. 27, then though there could be equilibrium without any transference of water from A to B, it is easy to see that the equilibrium would be unstable. For suppose a little water were to flow from A to B, this would make B heavier and it would descend; the water in B would now be at a lower level than that in A, so that the water instead of flowing back as it would if the equilibrium were stable would continue to flow into B, and the flow would go on until the pressure due to the compression of the air confined above B was sufficient to balance the pressure difference arising from the difference of level. Thus the coupling up of the two would produce a transference of water from the one to the other, or, if we suppose that the water represents an electric charge, the one would be positively, the other negatively, electrified.

In the case of groups of corpuscles we should have forces between the groups with somewhat the same properties as those discussed in the last example. Thus, take one of the arrangements discussed on page 117, say the group of 62: this is more stable than the group of 61 and less stable than that of 63, or, as we have expressed it, the corpuscular pressure in the group 62 is less than that in 61 and greater than that in 63. Now suppose two groups each containing 62 were brought near together, and suppose a corpuscle were transferred from one group to the other, so that one group contained 61 and the other 63, as the pressure in the group 61 is greater than that in the group containing 63, the corpuscles would tend to go on flowing from 61 to 63 instead of coming back, that is, if one got by chance a negative charge, that charge would tend to increase until the electrostatic repulsion due to the negative charge was sufficient to counterbalance the effect of corpuscular pressure. Thus we see in this case that the stable configuration for two groups placed within range of each other's action is one in which there is a positive charge on one group and a negative charge on the other. If we apply these considerations to the case of atoms, we arrive at the conclusion that when two atoms of the same kind come so near

130 THE CORPUSCULAR THEORY OF MATTER.

together as to exert appreciable forces on each other one of them may become positively, the other negatively, electrified. Thus the two atoms in a diatomic molecule of an elementary gas may be oppositely electrified, and the forces which hold two similar atoms together in the molecule of an elementary substance may be quite similar to those which hold together two dissimilar atoms in the molecule of a compound. The maximum charge which an atom could receive when in combination with an atom of the same kind would be the same as the maximum charge when combined with an atom of a different kind, and would be determined by its valency. We can, as the example on page 121 shows, conceive of attractions between atoms of the same kind even when the atoms do not get oppositely electrified; but the properties of molecules of simple and compound gases seem to testify in favour of the view that the forces which hold the similar atoms together in the molecule of an elementary gas are of the same character as those at work binding together dissimilar atoms in the molecule of a compound. Thus, for example, gases such as helium or argon, whose atoms do not combine with the atoms of other gases to form compounds, do not combine with each other to form diatomic molecules. Again, when, as in carbon compounds, we have atoms of the same kind in combination with each other, the bonds uniting a carbon atom to another carbon atom are treated as following the same laws as to valency as those which bind the carbon atoms to atoms of a different kind.

The view that the atoms in a molecule are oppositely charged receives support from some experiments made by Walden, in which it was found that electrolytic conduction took place when bromine and iodine were dissolved in certain solvents, the bromine or iodine appearing at both electrodes, the results being consistent with the view that the bromine or iodine molecules are dissociated into the ions Br_+, Br_- or I_+, I_-. The view is also supported by the fact that when the molecules of an elementary gas are dissociated by heat, as in the case of iodine vapour, the electric conductivity of the dissociated gas is very high, showing that there are large quantities of both positive and negative ions present in the dissociated gas.

The optical properties of gases, especially the refractive index and the dispersion, would, as we shall see, be largely influenced by opposite charges existing on the atoms in the molecule—in fact, we should expect the dispersion in a gas in which the two atoms in the molecule carry opposite charges would be of quite a different order from the dispersion of a gas whose molecules consist of uncharged atoms. The numerous experiments which have been made on the dispersion of gases do not afford any evidence of the existence of any wide divergence between the dispersion of compound and of elementary gases; hence we may conclude that if the atoms in the molecules of the compound gas are charged with electricity, the atoms of the molecules of elementary gases are also charged.

The positive charge on one atom and the negative charge on the other produces a difference between the atoms which might lead to a want of symmetry in compounds which from their formulæ appear perfectly symmetrical. Thus, to take an example, ethane is represented by the formula—

$$\begin{array}{cc} H & H \\ \diagdown & \diagup \\ H\!-\!C & C\!-\!H \\ \diagup & \diagdown \\ H & H \end{array}$$

but if we regard the coupling up of one carbon atom with another as accompanied by a transference of a corpuscle from the one atom to another, the two carbon atoms will not carry equal charges. If all the hydrogen atoms are charged with one unit of positive electricity one of the carbons will have a charge of four units of negative electricity while the other will only have two units; thus of the two systems CH_3, one will carry a positive charge, the other a negative one.

This would involve the possibility of two isomeric compounds of the composition $C_2 H_5 Cl$, one when the chlorine is attached to the carbon atom with the charge 4, the other when it is attached to the carbon atom with the charge 2. I am not aware that there is any evidence of the existence of isomeric forms of C_2H_5Cl; it might be expected that even if both were stable they would have very different degrees of stability. It must be remembered that in considering the possibility of the existence of isomers from purely geometrical considerations all questions as to stability are ignored, so that isomers which are indicated by geometry as possible may be dynamically unstable and thus incapable of preparation.

If we consider compounds in which the carbon atoms are linked by more bonds than one we see the possibility from geometrical considerations of isomers among the hydrocarbons themselves. Thus consider ethylene—

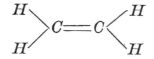

in which the carbon atoms are united by double bonds. If we regard each bond as involving the transference of a corpuscle from one carbon atom to the other, we might have two isomers; in one the transference of the corpuscle having been in the same direction along each bond, one carbon atom has lost and the other gained two units of negative electricity; in the other modification the transference of corpuscles has been in one direction along one bond and in the opposite direction along the other, so that on the whole the charge on the carbon atom has not been affected by the linkage. This form of compound is much more symmetrical than the preceding and will not give rise to so many isomers when chlorine is substituted for hydrogen. Again, even with single linkage between the carbon atoms we might have isomers among the hydro-carbons when the number of carbon

ARRANGEMENT OF CORPUSCLES IN ATOM. 133

atoms is greater than two. Thus consider the hydrocarbon represented by the formula— .

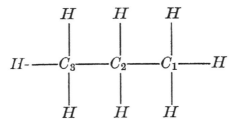

We might have one compound in which the linkage between C_1 C_2 caused a corpuscle to go from C_1 to C_2, and that between C_2 C_3 sent a corpuscle from C_2 to C_3; the result would be that C_3 is negatively and C_1 positively charged relatively to C_2. Or again, if a corpuscle went as before from C_1 to C_2, but the linkage between C_2 and C_3 caused a corpuscle to go from C_3 to C_2, instead of from C_2 to C_3, C_1 and C_3 would be both charged positively relatively to C_2, and this would differ from the preceding arrangement. We should get a third case if the linkage between the carbon atoms caused one corpuscle to go from C_2 to C_1 and another from C_2 to C_3; in this case both C_1 and C_3 would be charged negatively relative to C_2. We should of course get a larger number of isomers if we had a large number of carbon atoms.

We thus see that in the carbon compounds the charge carried by the carbon atom will depend upon whether the elements combined with the carbon are electro-positive or electro-negative with respect to that element. Thus in the compound—

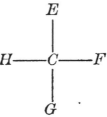

where C is the carbon atom and E F G H are monovalent atoms of other elements, if these elements are all electro-positive with regard to carbon, the carbon atom will carry a charge of 4 units of negative electricity, while if they are all electro-negative it will carry a charge of 4 units of positive electricity; if one is electro-positive the others electro-negative the charge on C will be 2 units of positive, and so on. Thus the properties of the carbon atom will depend upon the elements with which it is in combination. This variation in the properties might be difficult to detect in the saturated compound, but might be expected to exert more influence in organic radicals such as—

which enter into compounds by the linkage of their carbon atoms with other atoms; the facility with which this linkage takes place might be gravely affected by the sign and magnitude of the electric charge carried by the carbon atom, and this seems to be in accordance with the results of observations, as Van t'Hoff in his "Ansichten über Organischen Chemie" gives several instances of changes produced in the carbon atom in organic radicals by changes in the elements with which it is in combination.

A system of four atoms, each possessing unit positive and unit negative valency rigidly attached to each other and forming the four corners of a regular tetrahedron, would possess the same chemical properties as the carbon atom; two such atoms could be united by one, two, or three bonds, while the free valencies not satisfied by the connection between the atoms would be satisfied by any univalent atoms whether electro-positive or electro-negative.

We might also expect to see traces of the influence of the properties of the atom we have been considering on the boiling points of liquids or on the temperature at which gases are liquefied, as these depend on the forces exerted between different molecules of the substance, an increase

ARRANGEMENT OF CORPUSCLES IN ATOM. 135

in these forces tending to raise the boiling point of a liquid and increase the ease with which a gas is liquefied. These forces also influence the connection between the pressure and volume of a gas; they are, for example, responsible for the term a/v^2 in Van der Waals equation—

$$\left(p + \frac{a}{v^2}\right)(v - b) = R\theta,$$

and from the values of a we can deduce a measure of the intensity of these forces. It is more satisfactory to take a as the index of these forces than to use the boiling point or even the critical temperature for this purpose, as the latter depend upon b the size of the molecule as well as upon the inter-molecular forces. Another useful measure of the strength of their forces is the amount of heat required to turn one gramme molecule of the substance from the liquid into the gaseous state, as this quantity is directly proportional to the work required to drag away a molecule of compound against the attraction exerted upon it by a slab of the substance in the liquid state.

Let us now endeavour to picture to ourselves how these forces may arise. When an atom in a compound is "unsaturated" we should expect it to exert considerable attraction on other atoms, because we know that under suitable conditions it is able to attract some other atoms so firmly that they become permanently attached to it. But even when an atom in a molecule is "saturated," *i.e.*, when no transference of corpuscles to or from it is possible, forces between two neighbouring atoms may exist although they are not able to drag the corpuscles from one atom to another and thus establish a "chemical bond" between the atoms. The force between two atoms will, among other things, depend upon the ease with which the corpuscles can move about in the atoms, for just the same reason that the force between two oppositely electrified bodies is greater when these bodies are conductors of electricity in which the electricity can move about and electrostatic induction come into play, than when they are insulators in which the electricity is fixed.

If an atom is unsaturated, it means that there are still some corpuscles which have comparative freedom of motion, for under suitable conditions these can be moved into or out of the atom when it acquires its maximum valency; thus we should expect that a molecule containing an unsaturated atom would exert considerable forces upon other molecules, and thus would tend to make the gas depart from Boyle's law and to be easily liquefiable. But even when all the atoms in the molecule are saturated and the valency corpuscles transferred there may still remain some mobility of the corpuscles, although not sufficient to enable them to get clear of the atom; this mobility may not be the same for the atoms of the different elements, and may be different for the same atom according as it is exerting positive or negative valency; in other words, the attraction of an atom may not be wholly exhausted when its valency is satisfied, and the residual attraction may depend not only upon the nature of the atom but also upon whether it is exerting its positive or negative valencies.

Let us take some examples. Marsh gas CH_4 is a gas which is not easily liquefied and in which the attraction between the molecules is small; when, however, one of the hydrogen atoms is replaced by OH we get methyl alcohol, which is a liquid at ordinary temperatures and whose molecules exert considerable attractions on each other. If, as many chemists maintain, oxygen may be tetravalent, then the oxygen in CH_3OH is unsaturated, and can exert considerable attractions on other atoms; the hydrogen which it replaced was saturated, and could not therefore exert nearly such large attractions. Again, chlorine is very far from being a perfect gas, and the wide deviations it shows from Boyle's law indicates that the residual attractions between the molecules is very considerable. Chlorine seems to retain this residual attraction when in combination with other elements, for the result of replacing the hydrogen atoms in CH_4 by chlorine atoms as in the compounds CH_3Cl,

$C\,H_2\,Cl_2$, $C\,H\,Cl_3$, $C\,Cl_4$, is to produce substances which depart more and more from being perfect gases as their chlorine content is increased; indeed, the later ones are fluid at ordinary temperatures. Assuming that the hydrogen atom is positively, and the chlorine atom negatively, charged, the charge carried by the carbon atom varies from -4 in $C\,H_4$ to $+4$ in $C\,Cl_4$, and it is an interesting subject of inquiry whether the residual attraction of the carbon atom is affected by the charge; the residual attraction of chlorine, however, is so great that it would probably swamp the effect of the carbon. Since the residual attraction of hydrogen is very small we should have a better chance of detecting changes in the residual attraction of carbon if we worked with compounds containing nothing but hydrogen in addition to the carbon. A study of the values of a in Van der Waals' equation for such compounds as $C_2\,H_6$, $C_2\,H_4$, $C_2\,H_2$, in which the carbon atoms carry charges which vary from one compound to another, might throw some light on this question. In the compound $C\,H_4$ the carbon is supposed to carry a charge of -4, and in $C\,O$ (if the oxygen is tetravalent) a charge of $+4$; the value of a for $C\,H_4$ is ·0379, and that for $C\,O$ is less, viz., ·0284, although the residual attraction of oxygen is probably greater than that of hydrogen. This, as far as it goes, is in favour of the view that the residual attraction of carbon is greater when it is negatively than when it is positively charged.

Besides affecting the relation between the pressure and volume, the residual attraction has apparently a great effect upon the specific inductive capacity of the substance. Thus, for example, the liquids which contain the radicles $O\,H$, $N\,O_2$, $C\,O\,H$, have generally very large specific inductive capacities; and moreover, as Drude has shown, frequently show anomalous dispersion for electric waves whose wave lengths are enormously greater than the size of a molecule; this suggests that the large residual attraction between the molecules leads to the formation of aggregates containing a very large number of molecules, and that the exceptionally large value of the specific inductive capacity is due to the presence in these liquids of such aggregates.

On the view of chemical combination given above, the valency of an element depends upon the number of corpuscles which can be transferred to or from an atom of the element by the action of atoms of other elements. For each valency bond established between two atoms the transference of one corpuscle from the one atom to the other has taken place, the atom receiving the corpuscle acquiring a unit charge of negative electricity, the other by the loss of the corpuscle acquiring a unit charge of positive. This electrical process may be represented by the production of a unit tube of electric force between the two atoms, the tube starting from the positive and ending on the negative atom. In this way we can give a physical interpretation to the lines by which in graphical formulæ the chemists represent the valency bonds, these lines representing the tubes of force which stretch between the atoms connected by the bond. Thus, for example, the lines in the graphical formula—

represent the tubes of electric force which stretch between the carbon atom and the four hydrogen atoms. There is, however, one important difference between the lines representing the bonds and the tubes of electric force. The lines used by the chemist are not supposed to have direction. Thus, in the two compounds—

the lines joining the carbon atom to the hydrogen atom are not distinguished in any way from those joining the

ARRANGEMENT OF CORPUSCLES IN ATOM. 139

carbon to the chlorine atoms. On the electrical theory, however, the tubes of electric force are regarded as having direction starting from the positive and ending on the negative atom; thus, if the hydrogen atoms are positively electrified in marsh gas and the chlorine atoms negatively electrified in carbon tetrachloride, the graphical formulæ representing them would be—

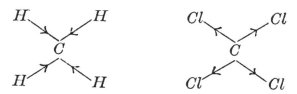

respectively, indicating that the carbon atom is not in the same condition in the two compounds, as in one case it is the terminus, in the other the origin of the tubes of electric force.

A method of investigating the magnitude and nature of the residual attraction exerted by a gas which seems not unlikely to lead to interesting results, is to find the effect produced on the velocity of the ions, positive as well as negative, through air when a small quantity of the gas under consideration is mixed with the air. Thus, to take an example of the effect we are considering, it has been found that in carefully dried gases the velocity of the negative ion is considerably greater than that of the positive when the electric forces acting on them are equal. If, however, a little water vapour is added to the gas it produces a considerable diminution in the velocity of the negative ion while it hardly affects that of the positive. It seems quite possible that this is due to the residual attraction of the OH radicle in the water for a negative charge, making the water molecules attract the negative ions more strongly than they do the positive ones, so that the water molecules will tend to attach themselves to the negative ions, and by loading them up diminish their velocity. It would be interesting to test this result by seeing whether other gases which contain the hydroxyl group OH possess, like water, the power of loading up negative ions to a greater extent than they do positive ones; whether also there are not other radicles or atoms which give to the compounds in which they occur the same property; and whether, though as yet undiscovered, there may not be other atoms or radicles which possess the power of loading up the positive ion more than the negative one.

Another property which may perhaps throw light on the different states of the two atoms in an elementary gas is the magnetic qualities possessed by some elements even when in the gaseous state. Perhaps one of the most interesting things in physics is the magnetic quality of oxygen. This gas when in the molecular condition is strongly magnetic; and ozone is even more magnetic than oxygen. Oxygen is so magnetic that liquid oxygen will fly to the poles of a bar magnet placed near it. But although oxygen is so magnetic in the molecular condition, it does not, with few exceptions, of which the most noticeable is nitric oxide (NO), preserve this quality in its compounds. Thus a mixture of hydrogen and oxygen in the proportion of two volumes of hydrogen to one of oxygen is magnetic, but if the hydrogen and oxygen, instead of being mechanically mixed, enter into chemical combination and form water vapour, the result is a diamagnetic substance. Again, equal volumes of oxygen and carbonic acid gas contain the same amount of oxygen, yet the oxygen is magnetic, the carbonic acid diamagnetic.

I am inclined to think that this property of oxygen is due to one of the atoms in a molecule of oxygen being in a state in which the atom is not usually found in oxygen compounds, and that it is to the atoms in this state that oxygen owes its magnetic quality. If we suppose that the two atoms in the molecule of oxygen are held together by electrical forces, one atom being positively, the other negatively electrified, then in the molecule of oxygen we have a positively electrified atom of oxygen, *i.e.*, an atom which has lost at least one, and probably more than one, corpuscle, while in at any rate the vast majority of oxygen

ARRANGEMENT OF CORPUSCLES IN ATOM. 141

compounds the oxygen atom is negatively electrified. Now, if we suppose that the oxygen atom which has lost corpuscles— *i.e.*, the positively electrified one—is magnetic, while the one which has gained them—*i.e.*, the negatively electrified atom— is either non-magnetic or diamagnetic, we can evidently easily account for oxygen being magnetic in O_2, and non-magnetic or diamagnetic in its compounds. In support of this view we may urge that a similar phenomenon seems to occur in the case of iron, for Townsend has shown that in solutions of iron salts the magnetic qualities of the iron depend greatly upon the nature of the salts: thus for all ferric salts the coefficient of magnetisation is equal to aN where N is the number of atoms of iron in unit volume of the solution, and a is a constant which does *not* depend upon the element with which the iron is in combination. Again, the coefficient of magnetisation of all ferrous salts is βN, where β is again independent of the other constituent of the salt; β is not the same as a; both the ferrous and ferric salts are strongly magnetic, and in them it will be noticed that the iron atom is positively charged. When, however, the iron atom occurs in the negative part of the molecule, as in the ferricyanides, Townsend proved that it showed no magnetic quality, the ferricyanides being no more magnetic than salts not containing iron. This shows that the iron atom may be magnetic or non-magnetic according as it is on the positively or negatively electrified side of the molecule; and the phenomena associated with the magnetism of oxygen and its compounds indicate that the oxygen atom possesses a similar property.

CHAPTER VII.

ON THE NUMBER OF CORPUSCLES IN AN ATOM.

If we take the view that corpuscles are an essential constituent of the atom, one of the most fundamental questions to be answered is, how many corpuscles are there in an atom?

We shall consider three methods by which an estimate of the number of corpuscles in an atom may be obtained.

METHOD I.—SECONDARY RONTGEN RADIATION.

This method is based on the determination of the proportion between the energy in a beam of primary Rontgen rays passing through a gas and that in the secondary Rontgen rays emitted by the gas through which the primary rays are passing.

When an electrified body is suddenly set in motion, or if when in rapid motion it is suddenly stopped, pulses of intense electric and magnetic force start from the body and travel outwards through space with the velocity of light. The pulses produced when the rapidly moving electrified particles in a vacuum tube, called cathode rays, strike against the walls of the tube and have their velocities rapidly diminished, constitute on our hypothesis the well-known Rontgen rays. We regard, then, the Rontgen rays as very thin pulses of intense electric and magnetic force. When such a pulse strikes against a corpuscle the electric force in the pulse acting on the electric charge on the corpuscle in a very short time makes the corpuscle move with a great velocity. This sudden starting of the electrified corpuscle produces another pulse of electric and magnetic force, and the aggregate of such pulses constitutes the secondary radiation emitted by the substance containing the corpuscles. The corpuscle will emit the pulse only while the velocity is changing, and if the corpuscle is free, or if the other forces acting upon it are small compared with those in the pulse of primary Rontgen radiation, the change in the velocity of the corpuscle will take place while, and only while, the primary pulse is passing over it, and the thickness of the secondary pulse will be the same as that of the primary; in this case the quality of the secondary radiation as measured by its penetrating power will be the same as that of the primary. If, however, the corpuscle, after being displaced by the primary ray, is acted upon by very intense forces due to the proximity of other corpuscles, it is evident that the character of the secondary pulse will not be the same as when those forces are small, for in the latter case the acceleration of the corpuscle will sink to a very small value as soon as the primary pulse has left the corpuscle, so that the primary and secondary pulses will have the same thickness, while in the former case the acceleration of the corpuscle will be large long after the pulse has passed the corpuscle; thus the breadth of the secondary pulse will be much increased, and the secondary radiation will be of a much more absorbable type than the primary. Again, if the primary pulse is so thick or the corpuscles are so closely packed that before the pulse has left one corpuscle it overlaps another, the pulses emitted by the corpuscles will not be separated by a finite interval, but will overlap and produce a pulse thicker than the primary one. Thus if we find that the secondary radiation is of the same type as the primary, we may conclude that the pulses are so thin that they act upon the corpuscles one at a time, and also that the forces of restitution called into play by the displacement of the corpuscle are small compared with the forces exerted on the corpuscle by the electric force in the primary pulse. When these conditions are fulfilled it can be shown (*see* J. J. Thomson, " Conduction of Electricity through Gases," 2nd edition, p. 326) that the energy of the secondary radiation emitted per unit time per unit volume of a space containing corpuscles is equal to $\dfrac{8\pi}{3}\dfrac{Ne^4}{m^2}$ times the energy in the primary radiation passing through unit volume in that time. N is the number of corpuscles per unit volume, e the charge on a corpuscle, and m its mass.

Now Barkla, who has made very valuable investigations on the Secondary Radiation produced by Rontgen rays, finds that for elements with small atomic weight the quality of the secondary radiation is the same as that of the primary. He finds, too, in accordance with the expression just found, that the proportion between the energy of the secondary radiation from such elements and the primary radiation is independent of the nature of the primary radiation, and that for different substances this ratio is directly proportional to the density of the substance. Since this ratio is equal to $\dfrac{8\pi}{3}\dfrac{Ne^4}{m^2}$, this result shows that the number of corpuscles in unit volume is proportional to the density of the substance, and since the density is equal to the number of atoms per unit volume multiplied by the atomic weight, it follows from this result that the number of corpuscles in an atom is proportional to the atomic weight of the substance, e.g., that the number of corpuscles in an atom of oxygen is sixteen times that in an atom of hydrogen, and so on.

Barkla finds that the energy of the secondary radiation from one cubic centimetre of air at atmospheric pressure is about ·00025 that of the primary radiation passing through it. Hence we have—

$$\dfrac{8\pi}{3}\dfrac{Ne^4}{m^2} = \cdot 00025.$$

Now $e = 1\cdot 2 \times 10^{-20}$ and $e/m = 1\cdot 7 \times 10^7$. Substituting these values, we find—

$$Ne = 10.$$

If n is the number of molecules in a cubic centimetre of air at atmospheric pressure and $0°$ C. we know—

$$ne = \cdot 4.$$

NUMBER OF CORPUSCLES IN AN ATOM.

Thus $N = 25n$, so that the molecules of air contain on the average about twenty-five corpuscles. The molecular weight of nitrogen is 28; this result suggests that the number of corpuscles in the atom of nitrogen is equal to the atomic weight of nitrogen. Since the energy scattered by different gases is proportional to the density of the gas, the number of corpuscles per unit volume must be also proportional to the density; hence if the number of corpuscles in an atom of any one substance is equal to the atomic weight of that substance, the number in an atom of any substance whatever must be equal to the atomic weight of that substance.

As this is a question of very great importance it is necessary to consider carefully the assumptions made in proving that the energy radiated per unit volume is equal to $\frac{8}{3} \frac{N e^4}{m v^2}$ times the energy in the primary rays. N is the number of corpuscles set in motion by the primary pulse; we have assumed that all the corpuscles are set in motion. It might be that some of the corpuscles are bound to the sphere of positive electrification by forces so strong that practically the corpuscles are rigidly attached to the atom and cannot move without dragging the whole atom with them; the acceleration of corpuscles such as these would be so small that they would give rise to very little radiation in comparison with the freely moving corpuscles, so that any method based on secondary radiation would not be able to detect these fixed corpuscles. We shall see, however, that such could be detected by our second method.

Another assumption made was that the pulse of Rontgen radiation was so thin as not to contain more than one corpuscle at once. In favour of this is the fact that the secondary radiation from light substances is of the same quality as the primary, whereas if the pulse covered several corpuscles at once, the thickness of the pulse of the secondary radiation would be greater than that of the primary pulse by the distance between two corpuscles if the pulse at each collision passed over two corpuscles before it got free from a corpuscle,

146 THE CORPUSCULAR THEORY OF MATTER.

by twice this distance if it passed over three, and so on. This would make a great difference in the quality of the secondary radiation if the thickness of the primary pulse were only a small multiple of the distance between two corpuscles. It would make comparatively little difference if the thickness of the pulse were much greater than the diameter of an atom; so that the identity of the secondary and primary radiation is not inconsistent with very thick pulses, although it is so with moderately thick ones. If the pulses are thicker than the atom, then all the corpuscles in an atom will be moving as if they were a single charged body, with a charge $p\,e$, and a mass $p\,m$ if p is the number of corpuscles in the atom; hence the energy radiated per atom will be $\frac{\pi}{3}\frac{p}{m}$; if n is the number of atoms per unit volume, the energy radiated per unit volume is $\frac{\pi}{3}\frac{e^4}{m^2}n\,p^2$. Since experiment shows that this is proportional to the density, i.e., to $n\,M$ if M is the atomic weight, we see that if the Röntgen radiation were of this character p^2 would be proportional to M, so that p, the number of corpuscles in the atom, would be proportional to the square root of the atomic weight.

The fact that the secondary radiation from light liquids and solids, as well as of gases, is of the same character as that of the primary, shows, however, that the pulses in the primary radiation cannot be so thick as we have supposed, for if the thickness of the pulse were much greater than the diameter of an atom then such a pulse on passing through a solid or liquid consisting of such atoms would never be free from corpuscles, and the secondary pulse would be very much thicker and more drawn out than the primary.

When the primary pulse is thicker than the atom and the electric force is in the same direction from back to front, the energy radiated per atom is proportional to the square of the number of corpuscles in the atom instead of to the first power of the number, as it is when the pulses are thin; thus the radiation increases more rapidly with the number of corpuscles for thick pulses of this type than for thin ones. This, however, is only true when the electric force is uniform in direction throughout the pulse. When the electric force in the pulse, instead of being uniformly in one direction, is in one direction in the front of the pulse and in the opposite direction in the rear, the radiation from a thick pulse will be less than from a thin one, for the corpuscles in the rear of the thick pulse will have opposite accelerations to those in the front; the electric and magnetic forces produced by them will be in opposite directions, they will neutralise each other's effects, and the radiation from the two combined will be much less than from either individually. On the other hand, when the pulse is so thin that it can only cover one corpuscle at a time the radiation from each corpuscle will spread out independently and the energy radiated will be directly proportional to the number of corpuscles.

The secondary radiation from substances with large atomic weights is not entirely of the same character as the primary radiation, indeed, from these substances by far the greater part of the secondary radiation consists of cathode rays of an easily absorbable type, so that as a whole the secondary radiation is very much less penetrating than the primary. Barkla found that the amount of secondary radiation of the same penetrating type as the primary produced by substances with large atomic weights is often less than that from the lighter elements; in the case of the latter, however, practically all the secondary radiation is of a penetrating type, while this type forms only a fraction of the total radiation from the heavier elements. We should expect the penetrating radiation to begin to diminish if the corpuscles in the atom get so crowded that the Rontgen pulses spread over more than one corpuscle at once, provided that the direction of the electric force in the pulse is reversed between the front and the rear of the pulse. Barkla found that with the rays he used, the elements of smaller atomic weight than calcium gave off secondary radiation of the same type as the primary, while the secondary radiation from calcium and elements with larger atomic weights consisted chiefly of easily absorbable cathode rays. I found that the element at which the change took place depended upon the quality of the primary rays, and that with very soft rays the change in the character of the secondary radiation can take place at lighter elements than calcium.

If, as the result of the passage of the primary rays through the gas, the gas is ionized, *i.e.*, if corpuscles are detached from the atoms, the collisions made by these corpuscles will increase the secondary radiation; this increase will, however, consist of a radiation which is not in general of the same type as the primary.

Second Method of Estimating the Number of Corpuscles in an Atom: by Determining the Opacity of a Substance to Cathode Rays.

If a cathode ray is travelling with a very high velocity through a collection of corpuscles, then when it passes close to one of the corpuscles it will be deflected; as the result of such deflections a bundle of cathode rays originally parallel to the axis of x will get more and more diffuse as they pass through the substance, and the number passing in unit time through a unit area at right angles to the axis of x will get smaller and smaller as the length of path of the rays increases. The amount of deflection experienced by the moving corpuscle will depend to some extent upon the firmness with which the corpuscles in the absorbing substance are held in their positions of equilibrium by the forces inside the atom. The solution of the problem when these forces are taken into account would be exceedingly difficult and complex. We may, however, represent to some extent the action of these forces by increasing the mass of the corpuscles in the absorbing substance, the effect of a corpuscle that is held absolutely rigidly by the forces acting upon it being the same as if it were free from such forces but had an infinite mass.

It can be shown that on these suppositions the number of corpuscles which pass through unit area at right angles to the x axis, at a distance x from the place where the stream of corpuscles enters the substance, is equal to $I_0 \epsilon^{-\lambda x}$, where I_0 is the number when $x = 0$ and—

$$\lambda = 4\pi N e^4 \left(\frac{V_0}{V}\right)^4 \frac{M_1 + M_2}{M_1 M_2^2} \log \left(\frac{a}{e^2}\left(\frac{V}{V_0}\right)^2 \frac{M_1 M_2}{M_1 + M_2} - 1\right).$$

(See J. J. Thomson, *Phil. Mag.*, June, 1906, "Conduction of Electricity through Gases," 2nd edition, p. 377.)

In this expression N is the number of corpuscles per unit volume of the absorbing substance, e the charge on a corpuscle in electro-magnetic units, M_1 the mass of a corpuscle in the atoms of the absorbing substance, M_2 that of the moving corpuscle, V the velocity of the moving corpuscle, V_0 the velocity of light, and a the distance between the corpuscles in the atoms of the absorbing substance; λ is the coefficient of absorption of the cathode rays in the substance. We can express the value of λ in terms of P, the number of corpuscles in an atom of the absorbing substance, for if d is the density of the substance, μ the mass of an atom, $P d = N\mu$, thus—

$$\frac{\lambda}{d} = 4\pi \frac{P e^4}{\mu} \cdot \left(\frac{V_0}{V}\right)^4 \frac{M_1 + M_2}{M_1 M_2^2} \log \left(\frac{a}{e^2}\left(\frac{V}{V_0}\right)^2 \frac{M_1 M_2}{M_1 + M_2} - 1\right).$$

Now $e/M_2 = 1.7 \times 10^7$

$e = 1.2 \times 10^{-20}$

$\frac{e}{\mu} = 10^4 / w$, where w is the atomic weight of the absorbing substance, and $V_0 = 3 \times 10^{10}$ is the velocity of light.

Substituting these values we find—

$$\frac{\lambda}{d} = \frac{4\pi P}{w} \frac{V_0^4}{V^4} \times \frac{1.05}{30} \frac{M_1 + M_2}{} \log \left(\frac{a}{e^2}\left(\frac{V}{V_0}\right)^2 \frac{M_1 \times M_2}{M_1 + M_2} - 1\right).$$

The absorption considered in this investigation is that due to the scattering of the corpuscles by collisions with other corpuscles; the change in the kinetic energy of the colliding corpuscle is neglected. This absorption is analogous to the scattering which takes place when a ray of light passes through a layer of powdered glass. As it is important to have a clear idea of what is meant by this coefficient we shall consider a special case. Suppose the incident corpuscles form a thin cylindrical beam $E\ F\ G\ H$; after passing through the absorbing substance this beam will be scattered, and its section will be much larger. The coefficient of absorption we have investigated is measured by the diminution in the energy passing across a section of $L\ M\ N\ P$, the prolongation of the incident beam, and not to the diminution in the total amount of energy passing through the plate; this will of course be much less than the diminution in the energy passing through a section of $L\ M\ N\ P$. Thus if the corpuscles originate from a radioactive substance placed in a metal tube $E\ F\ G\ H$, the coefficient of absorption of the plate is measured by the diminution in the number which pass through a tube $L\ M\ N\ P$, the prologation of the tube in which the radioactive substance is placed, and not by the diminution in the number of corpuscles which find their way through the plate. In the experiments hitherto made on the absorption of the β rays the quantity measured has been the ionisation produced by all the rays which emerge from the absorbing layer; this may explain the reason why the values of λ/d obtained by various physicists for the β rays given out by radio-active substances are much smaller than the values for rapidly-moving cathode rays produced in a vacuum tube, when the absorptions have been determined by measuring the quantity passing through a constant area, like that of the section of the tube $L\ M\ N\ P$, and not the quantity passing through the whole of the absorbing plate. That this difference is very marked and too great to be accounted for by differences in the velocities of the corpuscles may be seen from the fact that λ/d varies from 5 to 10, for the β rays from uranium, which, according to Becquerel, have a velocity of $1\cdot 6 \times 10^{10}$ cm/sec., while Becker found that for cathode rays with a velocity of 10^{10} cm/sec. λ/d varied between 1,200 and 2,000. Some experiments recently made at the Cavendish Laboratory by Mr. Crowther have shown that when λ is the diminution in the quantity of rays passing through unit area λ/d for rays from uranium is as much as 150. As the quantity measured in the experiment with the cathode rays corresponds to our coefficient of absorption, while that for the uranium rays does not, we shall use the former to determine the value of P/w.

Taking the case investigated by Becker, where $V = 10^{10}$, and putting $M_1 = M_2$, we find from the equation on page 149 that—

$$\frac{\lambda}{d} = 67 \frac{P}{w} \log\left(\frac{a}{9} \frac{M}{e^2} \frac{1}{2} - 1\right).$$

But (see p. 34) $M = \frac{2}{3} \frac{1}{b}$ where $b = 10^{-13}$ and is the radius of a corpuscle, hence—

$$\frac{\lambda}{d} = 67 \frac{P}{w} \log\left(\frac{a}{27 \times 10^{-13}} - 1\right)$$

if we take a to be of the order 10^{-8}

$$\frac{\lambda}{d} = 67 \frac{P}{w} \log\left(\frac{10^5}{27}\right)$$

approximately—

$$= 6 \frac{P}{w}$$

As λ/d varies between 1,200 and 2,000, we see from this that P/w cannot be large, i.e., that the number of corpuscles in the atom must be of the same order as the atomic weight.

This method supplements the preceding method (p. 142), for on the former method corpuscles which are so firmly held that they are not moved by the Röntgen rays would not be accounted for. The present method is not open to this objection; on the other hand, this method involves the assumption that, however small may be the distance between the corpuscles, the repulsion between them varies inversely as the square of the distance.

152 THE CORPUSCULAR THEORY OF MATTER.

The Information Given by the Optical Properties of Bodies on the Structure of the Atom.

We could estimate the number of corpuscles if we had measurements of the dispersion of light by a monatomic gas. For it can be shown (see *Phil. Mag.*, June, 1906) that if the atom contains n corpuscles in a sphere of uniform positive electrification, the refractive index μ of the gas for light waves with a frequency p is given by the equation—

$$\frac{\mu^2 - 1}{\mu^2 + 2} = \frac{\frac{4}{3}\pi N (m E^2 + M E e)}{\frac{4}{3}\pi \rho (M e + m E) - m M p^2} \quad (1)$$

here N is the number of atoms per cubic centimetre, e the charge on a corpuscle whose mass is m, M the mass of the sphere of positive electrification, E the magnitude of the positive charge, so that $E = n e$, ρ is the density of the positive electricity. The waves of light are supposed to be very much longer than the diameter of an atom, so that the electric force in the light wave is regarded as constant throughout an atom.

For infinitely long waves $p = o$, hence

$$\frac{\mu^2 - 1}{\mu^2 + 2} = \frac{N E}{\rho},$$

= N (volume of the sphere of positive electrification).
= volume occupied by these spheres per cubic centimetre of the gas.

This value agrees with that given by Mossotti's theory, where the atoms are regarded as perfectly conducting spheres.

If the term in p^2 is small, Equation (1) may be written—

$$\frac{\mu^2 - 1}{\mu^2 + 2} = \frac{N E}{\rho} \left\{ 1 + \frac{M m}{E e} \cdot \frac{3 E}{4 \pi \rho} \cdot \frac{1}{M + n m} p^2 \right\}$$

$$= \frac{N E}{\rho} \left\{ 1 + \frac{m}{e^2} \cdot \frac{3 E}{4 \pi \rho} \cdot \frac{M}{n (M + n m)} p^2 \right\}$$

the only factor which involves n is $\dfrac{M}{n (M + n m)}$ and this is always less than $1/n$, thus the dispersion (for the same sized atom) will diminish rapidly as n increases, and by measurements of the dispersion we could get an estimate of the value of n. Though we have at present no measurements of the dispersion of the monatomic gases, there seems reason from some experiments made by Lord Rayleigh to believe that it is of the same order as for diatomic gases. The dispersion of hydrogen has been shown by Ketteler to be given by the formula—

$$\frac{\mu^2 - 1}{\mu^2 + 2} = \frac{1}{3}\left(2\cdot 8014 \times 10^{-4} + 2 \times 10^{-14}\frac{1}{\lambda^2}\right)$$

where λ is the wave length.

Comparing this with the above formula, and remembering that $p = \dfrac{2\pi V}{\lambda}$ where V is the velocity of light, we find—

$$\frac{1}{n}\frac{M}{M + nm} = 1 \text{ approximately.}$$

This result shows that n cannot differ much from unity, hence if a monatomic gas had the same density and the same optical properties as hydrogen, it could not have many corpuscles in the atom. This result confirms those given by the preceding methods, that the number of corpuscles in the atom is proportional to the atomic weight.

The preceding expression for the refractive index of a gas involves the assumption that the force exerted by the positive electricity tending to bring a corpuscle back to its original position when displaced is equal to μ times the displacement where μ is the same for all the corpuscles. This assumption, however, is not true for diatomic molecules where the atoms are held together by the forces resulting from the displacement of the valency corpuscles. Thus, to take a simple case, when we have two atoms each containing one corpuscle, the atoms being of different sizes, when the smaller atom gets a certain distance inside the larger one, the corpuscle which was originally at the centre of the larger atom suddenly jumps into the smaller one and takes up a position at E inside the smaller atom, E being on the same side of B, the centre of the smaller atom, as A

the centre of the larger one. The corpuscle which was originally at B is displaced to F', F and E being on opposite sides of B. The corpuscle at E corresponds to the valency corpuscle.

Consider this corpuscle when it was in a position E_1 before moving into the smaller atom: if it is displaced through a distance ξ, the forces due to the positive electrification tending to bring it back to its original position are $\left(\dfrac{e^2}{a^3} - \dfrac{2\,e^2}{R\,F^3}\right)\xi$, while if F_1 was the corresponding position of F', the forces tending to bring it back are $\left(\dfrac{e^2}{a^3} + \dfrac{e^2}{b^3}\right)\xi$, where a and b are the radii of the larger and smaller atoms respectively. The coefficients of ξ in these expressions are different, and the preceding investigation does not apply. To find the expression for the refractive index in a case like this we may make use of a theorem due to Lorentz, which states that the refractive index μ for light whose frequency is p, due to a system of electrified particles with a charge e and mass m, and having $p_1, p_2, p_3, \ldots$ for the frequencies of their vibrations about their positions of equilibrium, is given by the equation—

$$\frac{\mu^2 - 1}{\mu^2 + 2} = \left(\frac{N_1 \dfrac{e^2}{m}}{p_1^2 - p^2} + \frac{N_2 \dfrac{e^2}{m}}{p_2^2 - p^2} + \ldots \right)$$

where N_1 is the number of systems per unit volume having the frequency p_1, N_2 the number having the frequency p_2, and so on.

It follows from this expression that the systems for which p_r is small make the largest contributions to the value of $(\mu^2 - 1)/(\mu^2 + 2)$. When p_r is small the force of restitution tending to bring the electrified particle back to its original position when it is displaced from it, is small, so that the particles which are easily displaced are those which have the greatest influence on the refraction. If we suppose that there are some particles which are so much more easily displaced than others that their influence on the refractive index swamps that of the other particles, and if we suppose that these loosely held particles have all the same period p_0, then we shall have, if N is the number of these particles per unit volume—

$$\frac{\mu^2-1}{\mu^2+2} = \frac{\frac{Ne^2}{m}}{p_0^2-p^2}$$

Now there is a number of substances for which the relation between the refractive index and the frequency can be expressed by the simple relation—

$$\frac{\mu^2-1}{\mu^2+2} = \frac{A}{p_0^2-p^2}$$

and the experiments of Ketteler and others furnish us with the values of A. Comparing this with the preceding expression for $(\mu^2-1)/(\mu^2+2)$ we see that—

$$A = \frac{Ne^2}{m}$$

Now since we know e/m and e, we can from this equation determine N, the number of these systems in unit volume. This has been done by Drude, who finds that the number determined in this way is greater than the number of atoms in that volume but not very much greater; it is very seldom, for example, as much as ten times greater. Also Drude found that the greater the chemical valency of the atoms in the refracting substance the greater was the number of the refracting systems per atom. Since in substances of great atomic weight the number of these refracting systems is only three or four times the number of the atoms, and therefore small compared with the number of corpuscles, it is evident that the corpuscles in the atom do not all take the same share in producing refraction, but that practically the whole of the work is done by a small fraction of the corpuscles, and that the greater the valency the larger is the number of corpuscles which give rise to refraction. We should expect a result of this kind, for the valency gives an indication of the number of corpuscles displaced, and these corpuscles, which are displaced by the action of one atom on others, are in all probability much less firmly fixed than those which retain their positions under this action; thus since the valency corpuscles are the ones most easily moved they are the ones which produce the greatest effect upon the refractive index. The optical properties of other than monatomic gases are thus complicated by considerations which make them unsuitable for determining the total number of corpuscles in the atom.

In a discussion of the optical properties of gases we must consider a very obvious objection to the view that the number of corpuscles in the atom is not a large multiple of the atomic weight. The objection is as follows: If the lines in the spectrum are due to the vibrations of the corpuscles in the atom, then since for n corpuscles there are $3n$ degrees of freedom, the maximum number of different periods of vibration of the system, *i.e.*, of lines in the spectrum, is $3n$. Hence on this view the number of corpuscles in the atom could not be less than one-third the number of lines in the spectrum, and this would for many elements be very much greater than the number representing the atomic weight. The case is, however, even stronger than this, for all, or nearly all, the lines in the line spectra of the elements show the Zeeman effect, *i.e.*, they can be resolved by a magnetic field into at least three components; thus each line showing this effect must correspond, not to a single isolated period, but to the coalescence of three equal periods. Now if we consider the theory of the vibrations of a system of n corpuscles, we find that p^2, where p is the frequency of a vibration, is given by an equation of the $3n$ degree, which could have at most $3n$ different roots. Many of these roots, however, would be isolated and the lines in the spectrum corresponding to them would not show the Zeeman effect; it is only the comparatively small number of frequencies, for which three of the roots of the equation are equal, which would give rise to lines having the properties of the lines in the spectrum. Thus if the spectrum of a body arose from the vibrations of the corpuscles in the atoms the number of corpuscles would have to be very greatly in excess of the number of lines in the spectrum, and therefore very much larger than the atomic weight.

I would urge against this objection that we have no evidence that the majority of the lines in the spectrum of an element arise from the atoms when in the normal state. Luminosity occurs when a gas is either traversed by an electric current or when it is raised to a high temperature, and in either case the gas is ionized, *i.e.*, we have, in addition to the normal atoms, positively charged ions and negatively electrified corpuscles. A positively electrified ion and a corpuscle might form a system analogous to the solar system, in which the positively electrified ion, with its large mass, takes the part of the sun while the corpuscles circulate round it as planets. The forces acting on the corpuscles are in part due to the attraction of the positive charge which produces a force varying inversely as the square of the distance, and in part due to the forces arising from the corpuscles and positive electricity within the ion. Thus, except in the case where there is a single corpuscle at the centre of a sphere of positive electrification, these forces will be finite and will vary not only with the distance of the corpuscle from the ion but also with the angular position of the corpuscle.

Now the question arises whether such a system could give rise to vibrations having definite periods separated by finite intervals as in the case of the line spectrum of the gas. In order that the corpuscle outside the ion may give a definite line it must revolve in a closed orbit; if orbits having all possible periods within certain limits were possible, then the systems of ions and corpuscles would give a continuous and not a line spectrum. Now if the forces between the positive ion and the corpuscle were simply a central force varying inversely as the square of the distance, there would be an infinite number of elliptic orbits for the corpuscle with continuously varying periods, and the spectrum would be a continuous one. When, however, as in our case, the force between the ion and the corpuscle is much more complex, the number of possible periodic orbits becomes much more limited. For some laws of force no periodic orbits at all exist; this, for example, is the case where the total force on the corpuscle is that due to a simple electrical doublet.

The results obtained by Sir George Darwin in his paper on "Periodic Orbits" in the *Acta-Mathematica* have an important bearing on the question we are discussing. Darwin discusses the possible periodic orbits of a particle of infinitesimal mass under the action of the sun and a planet

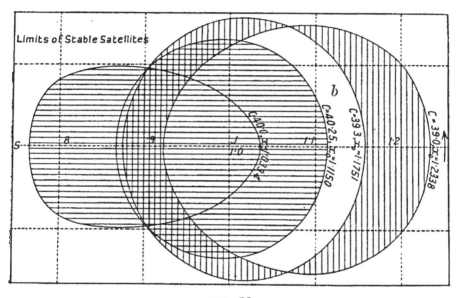

FIG. 28.

whose mass is 1/10 that of the sun. If we consider the orbits in which the particle moves as a satellite round the planet, the forces acting on the particle will consist of a central force varying inversely as the square of the distance and a force due to the attraction of the sun. We can resolve this force into two forces, one acting towards the planet and the other at right angles to the line joining the planet with its satellite, and we get some resemblance between these forces and those on a corpuscle arising from a very simple atom with a positive charge. The radial attraction due to the positive charge corresponds to the attraction of the planet,

NUMBER OF CORPUSCLES IN AN ATOM.

while the forces due to the corpuscles and sphere of positive electrification, though probably much more complex, may be compared with the attraction of the sun.

Now Darwin found that in the neighbourhood of the planet there is a region (unshaded area b, Fig. 28) through which no periodic orbit can pass, and though with the proportion between the masses of the sun and planet of 10 to 1, the region is not closed, Darwin expresses the opinion that with a larger value of this ratio this space would extend to an annular ring round the planet. We may, perhaps, imitate the greater complexity of the forces exerted by the atom over those at work in Darwin's problem by supposing that not one but several suns disturb the motion of the satellite; when it seems not improbable that we might have several rings instead of one in which periodic orbits are impossible. If these rings grow we might get to a condition of things in which the paths of periodic orbits are confined to a number of annuli (1), (2)...(n), between these rings. If the times of describing the orbits in region (1) vary from T_1 to $T_1 + \Delta T_1$, those in the region (2) from T_2 to $T_2 + \Delta T_2$, and so on, the lines in the spectrum given out by systems formed by the positive ion and one corpuscle, would consist of a line of finite width corresponding to periods of vibration varying from T_1 to $T_1 + \Delta T_1$, followed by another line with periods from T_2 to $T_2 + \Delta T_2$, and so on: if $\Delta T_1/T_1$, $\Delta T_2/T_2$ are small these lines will be sharp, while if these quantities are appreciable the lines will be broad. On this view the different lines are given out by different systems, the line T_1 by a positive ion and a corpuscle travelling round the region (1), the line T_2 by a positive ion and a corpuscle travelling round the region (2). If there were two corpuscles travelling round the same ion, one corpuscle being in region (1) the other in region (2), the two corpuscles would repel each other and soon make the orbits very irregular. The very large variations that take place in the relative intensities of different lines in the spectra produced by electrical discharges, by slight alterations in the discharge seem in accordance with the view that the different lines

160   THE CORPUSCULAR THEORY OF MATTER.

originate in different systems. The periods $T_1\, T_2 \ldots$ are determinate if we know the law of force exerted by the ion; the values will be connected with each other by certain relations: in other words, the vibrations corresponding to $T_1, T_2 \ldots$ would form what is called in spectroscopy a series. If we had an ion with a charge of two units of electricity instead of one,

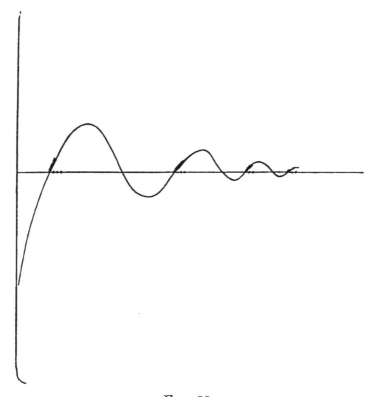

Fig. 29.

the regions (1) (2)…would be displaced and the times T_1, $T_2, \ldots$ altered, so that we should get a new series.

Any line arising from the revolution of a corpuscle in a closed orbit would show the Zeeman effect.

Another way of regarding the problem which leads to similar results is as follows:—Suppose we regard the charged ion as a Boscovichian atom exerting a central force on a corpuscle which changes from repulsion to attraction and from attraction to repulsion several times between the surface of the ion and a point at a distance from the surface comparable with molecular distance, such a force, for example, as is represented graphically in Fig. (29) where the abscissæ represent distances from the atom, and the ordinates the forces exerted by the atom on a corpuscle at a distance represented by the abscissa, the forces being repulsions when the representative point is below the line, attractions when it is above it. Now from any point where the force is attractive it is possible to project a corpuscle at right angles to the radius with such a velocity that it will, if free from disturbance, describe a circular orbit round the atom. The theory of central orbits shows, however, that only under a certain condition would these orbits be stable and able to exist in a system like that in a luminous gas subject to external disturbances. The condition that the circular orbit should be stable when its radius is a is that if P, the central attractive force due to the atom at a distance r, is written in the form $P = u^2 \phi(u)$, where $u = 1/r$, then

$$\frac{a\, d\, \phi(a)}{\phi(a)\, d\, a} \text{ must be less than unity.}$$

Now this condition will only be satisfied at parts of the Boscovichian curve; if these parts are denoted by the thickened portions of the curve in Fig. 29, the possible orbits will be confined to distances from the atom corresponding to the dotted belts in the figure, and we shall get the same conditions as before and the same arguments will apply.

It might be urged against the view that the vibrations in the line spectra of the elements are due to systems manufactured in the flame or in the electric discharge, and which do not exist in the normal atom, that the reversals of the bright lines in a spectrum show that in the reversing layer there are systems which have the same periods as those producing the lines, and therefore if the reversing layer consists of gas in its normal state there must in such a gas be systems having the same periods of vibration as the lines in the spectrum. It must, however, be remembered that in at any rate the great majority of cases the reversing layer does not consist of gas in its normal state; this layer is in the immediate neighbourhood of the luminous gas in the arc, spark or flame, or is itself at a high temperature. In all these cases it is ionised, *i.e.*, contains positive ions and corpuscles which may build up a system like those which we have supposed to be the origin of the bright lines, and absorb the light having the same period as those lines.

On the Origin of the Mass of the Atom.

Since the mass of a corpuscle is only about one-seventeen-hundredth part of that of an atom of hydrogen, it follows that if there are only a few corpuscles in the hydrogen atom the mass of the atom must in the main be due to its other constituent—the positive electricity. Now we have seen that the mass of the corpuscle may be regarded as arising wholly from its charge, and it might appear that this result obliged us to regard mass as arising in two distinct ways, the origin of one kind of mass—that of the corpuscles—being electrical, while that of the rest of the atom is mechanical. It is, however, I think, possible to take a point of view from which this separation of the nature of these two masses disappears. In my "Electricity and Matter" (p. 6) I showed that we might regard the mass of a corpuscle as the mass of the ether carried along by the tubes of electric force attached to the corpuscle as they move through the ether. An example taken from vortex motion through a fluid may make this idea clearer. When a vortex ring moves through the fluid it carries along with it a volume of the fluid which may be very much greater than the volume of the ring itself; in fact, if the ring is very thin and the velocity very great, the volume of the ring will be quite insignificant with that of the fluid which it carries along with it. Now the effective mass of the ring will be the mass of the ring itself *plus* the mass of the fluid it carries with it, and when the ring is thin the effective mass will be practically that of the attached fluid. The ring is a closed curve without ends. Let us consider the case of a vortex filament which is not closed but which has ends. The theory of vortex motion teaches us that these ends must, if they are not on the free surface of the liquid, be on bodies or cavities in the liquid. Let us suppose that the ends are on two bodies, A and B, which are so light as to have no appreciable mass of their own. Now when the system consisting of A and B and the connecting vortex filament moves through the fluid it will carry with it a certain volume of the fluid, and if the filament is very thin the effective mass of the system will be the mass of this fluid carried along with it by the system. Now this fluid is carried (1) by the vortex filament; (2) by the bodies A and B; if, for example, the latter are spheres, they will each carry along with them a volume of the fluid amounting to one-half of their own volume. Let us compare this system with that of a unit of positive electricity connected by tubes of electric force with a unit of negative electricity, the tubes of electric force corresponding to the vortex filament and the seat of the positive and negative electrification to the bodies A and B. We may suppose that when this system moves through the ether it carries some of the ether along with it; the portion carried by the tubes of force will depend on the distribution of these tubes, and since this distribution depends on the velocity, the mass of the ether carried along in this way will depend upon the velocity; the portions of the ether carried by the seats of the electrification will, if our analogy holds, not depend upon the velocity. We may interpret the results of the experiments described on page 33 as indicating that the amount of the ether carried by the seat of the negative electrification is very small compared with that carried by the tubes of electric force, and the result that the mass of the positive electricity is large compared with that of the corpuscle as indicating that the amount of ether carried along by the seat of the positive electricity is very large compared with that carried by the tubes of electric force and the seat

164 THE CORPUSCULAR THEORY OF MATTER.

of the negative electricity; that, in fact, the system of the positive and negative units of electricity is analogous to a large sphere connected with vortex filaments with a very small one, the large sphere corresponding to the positive electrification, the small one to the negative.

On the Size of the Sphere of Positive Electrification.

The connection between the volume of the sphere of positive electrification and the number of corpuscles in the atom is a very important question on the theory of the structure of the atom which we have been discussing. The number of corpuscles in the atom is equal to the number of units of positive electricity in the sphere, and is proportional to the atomic weight.

The great majority of the methods by which the size of atoms is determined do not give the geometrical boundary of the atom, but what is called the range of molecular action, *i.e.*, the greatest distance at which the forces due to the atom produce appreciable effect; they give, in fact, the dynamical rather than the geometrical boundary of the atom. On a theory such as that of Boscovich, in which the atoms are regarded merely as centres of force, the dynamical boundary is the only one which has to be considered; but in a theory such as the one we have been discussing, where we regard the atom as having a definite size and shape, we have to consider the geometrical as well as the dynamical boundary of the atom.

There is one method, however, by which we can in certain cases deduce the geometrical boundary of the atom, for we have seen that for a monatomic gas, if μ is the refractive index for infinitely long waves—

$$\frac{\mu^2 - 1}{\mu^2 + 2} = N a^3$$

where a is the radius of the sphere of positive electrification and N the number of atoms per unit volume of the gas.

NUMBER OF CORPUSCLES IN AN ATOM.

For a gas μ is so nearly equal to unity that we may write $\frac{1}{2}(\mu - 1)$ for $\frac{\mu^2-1}{\mu^2+1}$, so that for gases $\mu - 1$ will be proportional to the volume of the sphere of positive electrification. The following Table, the data for which are taken from the paper by Cuthbertson and Metcalfe (*Phil. Trans.*, A., vol. 207, p. 138, 1907), gives the value of $\mu - 1$ for many of the elements when in the gaseous state:—

Gas.	$\mu - 1$.	Atomic weight.	$10^6 \times (\mu - 1) /$ (atomic weight).
Helium	72×10^{-6}	4	18
Neon	137×10^{-6}	20	6·85
Argon	508×10^{-6}	40	12·7
Krypton	850×10^{-6}	80	10·6
Xenon	1378×10^{-6}	128	10·7
Mercury	1866×10^{-6}	200	9·3
Hydrogen	139×10^{-6}	1	139
⎧ Nitrogen	297×10^{-6}	14	21
⎨ Phosphorus	1197×10^{-6}	31	39
⎩ Arsenic	1550×10^{-6}	75	20
⎧ Oxygen	270×10^{-6}	16	17
⎪ Sulphur	1101×10^{-6}	32	34
⎨ Selenium	1565×10^{-6}	79	20
⎩ Tellurium	2495×10^{-6}	127	20
⎧ Zinc	2060×10^{-6}	65	30
⎩ Cadmium	2675×10^{-6}	112	24

For the lighter elements the variations in $(\mu - 1)/$ (atomic weight) are very irregular, but for those of large atomic weight in one group this quantity is fairly constant, indicating that the volume of the sphere of positive electrification is roughly proportional to the atomic weight when there are a great many corpuscles in the atom.

In many compounds of the lighter elements the value of

$(\mu-1)$ does not increase nearly as rapidly as the density, and for a considerable number of such compounds the value of $(\mu-1)$ at constant temperature and pressure is, as Traube has shown, approximately proportional to the sum of the valencies of the atoms in a molecule of the compound. The preceding table shows that this result does not apply to the heavier elements.

We may illustrate the effect of valency on the refractive index in the following way :—We have supposed that there are in the atom some corpuscles equal in number to the valency which are especially easily moved. To represent the mobility of these corpuscles let us suppose that they are placed in a shell of positive electricity of small density around the much denser core which contains the rest of the corpuscles and the equivalent quantity of positive electricity. Thus we may picture the atom as having a crowded centre, surrounded by a rarified atmosphere through which a few corpuscles are scattered, the positive electricity in the atmosphere being equivalent to the negative charge on the corpuscles scattered through it.

Now we can easily see that the value of $(\mu-1)$ for a collection of atoms of this kind will consist of two terms, one proportional to the volume of the atmosphere, and the other proportional to the volume of the core. The volume of the atmosphere will be proportional to the number of corpuscles in it, *i.e.*, to the positive valency, while the volume of the core will be proportional to the number of the remaining corpuscles; this for elements whose atomic weight is large compared with their valency, will be proportional to the atomic weight.

Thus, the value of $(\mu-1)$ will consist of two terms, one proportional to the valency, the other to the atomic weight. When the atomic weight is not great the first term may be the important one, while for the heavier elements the effect of the atomic weight may overpower that of the valency.

The dispersion of the substance is influenced by the valency atoms to an even greater extent than the refractivity for we can show that μ, the refractive index for waves of length λ, is given by the equation

$$\frac{\mu^2-1}{\mu^2+2} = P_0 + P_0^2 \frac{m}{Ne^2} \frac{1}{n} \frac{3\pi}{\lambda^2}$$
$$+ Q_0 + Q_0^3 \frac{m}{Ne^2} \frac{1}{p} \frac{3\pi}{\lambda^2}$$

where N is the number of atoms per cubic centimetre, m and e the mass and charge of a corpuscle, P_0 the part of $(\mu^2-1)/(\mu^2+2)$ due to the core for infinitely long waves, Q_0 the part due to the atmosphere, n the number of corpuscles in the core, p the number of valency corpuscles. Since p is in general small compared with n, we see that unless P_0 is large compared with Q_0, the part of the co-efficient of $1/\lambda^2$ which depends on Q_0 will be much larger than that which depends on P_0, i.e., the dispersion will depend chiefly on the valency atoms.

As these valency atoms are easily detached we should expect that they would increase the amount of ionization when the gas is ionized by some external means. Bragg finds that the number of ions produced by the α rays from radium in equal volumes of different gases at the same temperature and pressure is proportional to the molecular volume of the gas. As this molecular volume is proportional to $(\mu-1)$, and as the valency corpuscles have great influence on this quantity for the lighter elements, we see that they also increase the ionization.

INDEX.

A.

α PARTICLES, properties of, 25
    value of e/m for, 24
Abegg, on valency, 118
Absorption of cathode rays, 148, 150
Alloys, thermal and electrical conductivity of, 58, 59
Anomalous dispersion, 137
Arrangement of corpuscles in atom, 103 *et seq.*
Atom, Boscovichian, 160
    number of corpuscles in, 142 *et seq.*
    origin of mass of, 162
    volume of, 164
Atoms, forces between, 120
    forces between like atoms in a molecule, 127
Attraction, residual, 137

B.

BARKLA, energy of secondary radiation, 144, 147
Becker, absorption of cathode rays, 150
Becquerel, velocity of β rays from uranium, 150
Bonds, chemical, 136
Boscovichian atom, 160
Bose, effect of electric charge on electric resistance of a metal, 83
Boyle's law, interpretation of deviations from, 136
Bragg, properties of α particles, 25

C.

CANALSTRAHLEN, 17
    spectrum produced by, 18
    value of e/m for, 18 *et seq.*
Carbon atom, 133, 134
Cathode rays,
    absorption of, 148, 150
    determination of velocity of, 8
    electrostatic deflection of, 5
    magnetic deflection of, 7
    penetration of, Hertz experiments, 7
Charge of negative electricity carried by cathode rays, 4
Chemical combination, 120
Combination, chemical, 120
Condensation of water drops, 11
Conduction, electric, corpuscular theory of, 49, 86
Conduction metallic, corpuscular theory of, 49 *et seq.*
Conduction thermal, corpuscular theory of, 55, 88
Corpuscle, electric field due to, 44
    magnetic field due to, 44
    origin of mass of, 28
Corpuscles,
    arrangement in one plane, 107
    definition of 2
    electric charge carried by, 11
    mass of, 16
    number in an atom, 142 *et seq.*
    number of per unit volume of a metal, 80
    occurrence of, 10
Corpuscular pressure, 119

INDEX.

Corpuscular theory, statement of, 1
theory of radiation, 61 *et seq.*
Crookes, Sir W., experiment with cathode rays, 3
Cuthbertson and Metcalfe, refractivity of gases, 165

D.

DARWIN, SIR G., periodic orbits, 158
Deflection of cathode rays
by an electric field, 5
by a magnetic field, 7
Des Coudres, value of e/m for α particles, 24
Dewar and Fleming, effect of temperature on resistance of metals and alloys, 59
Diesselhorst and Jaeger, electric and thermal conductivities, 57
Dispersion, anomalous, 137
Dispersion of light, 152
influence of valency on, 155
Drude, anomalous dispersion, 137
effect of valency on dispersion, 155

E.

ELECTRIC charge carried by a corpuscle, 11
conduction, corpuscular theory of, 49, 86
field due to corpuscles, 44
force, tubes of, 138
Electricity, one fluid theory of, 26
Electrolysis of bromine and iodine solutions, 130
Electropositive and electronegative elements, 114
valency, 117, 122
Electrostatic deflection of cathode rays, 5
e/m values for α particles, 24

e/m values for canalstrahlen, 18 *et seq.*
corpuscles, 32
Ethane, 131
Ethylene, 132

F.

FITZGERALD, Peltier effect, 76
Fleming and Dewar, effect of temperature on resistance of metals and alloys, 59
Floating magnets, arrangement of, 110
Force, tubes of, 138
Forces between atoms, 120
between like atoms in molecule, 127
Fourier, analysis of radiation, 65, 91

G.

GASES, refractivity of, 165
Goldstein, canalstrahlen, 17

H.

HAGEN and Rubens, electric conductivity of metals, 84
Hall effect, 68, 99
Helium, canalstrahlen in, 21
Hertz, penetration of cathode rays, 7
Huff, value of e/m for α particles, 24
Hydrogen, canalstrahlen in, 21
dispersion of, 153

I.

IONS, velocity of, 139
Iron salts, magnetism of, 141
Isomers, 131

J.

JAEGER and Diesselhorst, electric and thermal conductivities, 57

INDEX. 171

K.

KAUFMANN, values of e/m for rapidly moving corpuscles, 32
Ketteler, formula for dispersion, 153
Kelvin, forces between electrified systems, 121
Kleeman, properties of a particles, 25

L.

LARMOR, radiation from moving corpuscle, 91
Light, dispersion of, 153
Lorentz, formula for dispersion of light, 153
    theory of radiation, 61
    Zeeman effect, 35

M.

MACKENZIE, value of e/m for a particles, 24
Magnetic deflection of cathode rays, 7
    force due to moving corpuscles. 43
    effect of, on flow of a current, 68
Magnetism of iron salts, 141
    oxygen, 140
Magnets, arrangement of floating, 110
Mass of atom, origin of, 162
    corpuscle, origin of, 28
Mayer, experiments with floating magnets, 110
Mercury vapour, conductivity of, 50
Minarelli, thermo-electric effects, 76
Metcalfe and Cuthbertson, refractivity of gases, 165
Molecule, forces between atoms in, 127
Monckman, experiments with floating electrified bodies, 112

N.

NUMBER of corpuscles in an atom, 142 *et seq.*
    per unit volume of a metal, 80

O.

OBERMAYER, thermoelectric effects, 76
Orbits periodic, 157
Origin of spectra, 157
Oxygen, magnetism of, 140

P.

PELTIER difference of potential, 73, 97
Periodic orbits, 157
Perrin, negative charge carried by cathode rays,
Positive electricity, 17
    from hot wires, 23
    from radioactive substances, 24
    size of sphere of, 164
Pressure, corpuscular, 119
Pulses produced by stopping and starting corpuscles, 45

R.

RADIATION, corpuscular theory of, 61
Radicles organic, 134
Rayleigh (Lord), 92
    conductivity of alloys, 59
Refractivity of gases, 165
Resistance, effect of magnetic field on electrical, 101
Reversal of lines in spectra, 161
Rontgen radiation, secondary, 142
Rontgen rays, theory of, 47

Rubens and Hagen, conductivity of metals, 84
Rutherford, value of e/m for α particles, 24

S.

SATURATED compounds, 135
Schulze, thermal and electric conductivity, 58
Secondary Röntgen radiation, 142
Size of sphere of positive electrification, 164
Spectrum, origin of, 157
    produced by canalstrahlen, 18
    reversal of lines in, 161
Sphere of uniform positive electrification, 164
Stokes (Sir G. G.),
    energy due to a sphere moving in water, 29
    velocity of falling drops, 14
Strutt, conductivity of mercury vapour, 50

T.

THERMAL conductivity, corpuscular theory of, 55, 88
Thomson effect, 76, 97
Townsend, magnetism of iron salts, 141
Traube, volume and valency, 166
Tubes of electric force, 138

U.

UNSATURATED compounds, 126, 135

V.

VALENCY, 115 *et seq.*
    corpuscles, 166
    positive and negative, 117
Van der Waals equation, 135
Vant Hoff, organic radicles, 134
Velocity of ions, 139
Volume of atom, 164
Vortex atom theory of matter, 2
    motion, analogy with electrical system, 162

W.

WALDEN, electrolysis of bromine and iodine solutions, 130
    on valency, 118
Wehnelt, lime coated cathode, 6
Wien W., value of e/m for canalstrahlen, 18
Wilson, C. T. R., condensation of water drops, 11
Wilson, H. A., measurement of charge on drops, 15
Wood, experiments with floating magnets, 112
Work required to disintegrate atom, 104

Z.

ZEEMAN effect, 34, 156

RETURN TO: CIRCULATION DEPARTMENT
198 Main Stacks

LOAN PERIOD 1 Home Use	2	3
4	5	6

ALL BOOKS MAY BE RECALLED AFTER 7 DAYS.

Renewals and Recharges may be made 4 days prior to the due date. Books may be renewed by calling 642-3405.

DUE AS STAMPED BELOW.

AUG 0 8 2006

FORM NO. DD6
50 M 1-06

UNIVERSITY OF CALIFORNIA, BERKELEY
Berkeley, California 94720–6000

PHYSICS LIBRARY

C064767919

MAR 1 * 2000

BOOK FOR PHYS

Made in the USA
Middletown, DE
30 November 2022